WALKING ON LAVA

WALKING ON LAVA
Selected Works for Uncivilised Times

THE DARK MOUNTAIN PROJECT

Edited by
Charlotte DuCann
Dougald Hine
Nick Hunt and
Paul Kingsnorth

Chelsea Green Publishing
White River Junction, Vermont

Project Manager: Alexander Bullett
Project Editor: Brianne Goodspeed
Proofreader: XXXXXXXXXXXXXXXXX
Designer: Melissa Jacobson

Printed in XXXXXXXXXXXXXXXXXXX.
First printing XXXX, 2017.
10 9 8 7 6 5 4 3 2 1   17 18 19 20 21

Our Commitment to Green Publishing
Chelsea Green sees publishing as a tool for cultural change and ecological stewardship. We strive to align our book manufacturing practices with our editorial mission and to reduce the impact of our business enterprise in the environment. We print our books and catalogs on chlorine-free recycled paper, using vegetable-based inks whenever possible. This book may cost slightly more because it was printed on paper that contains recycled fiber, and we hope you'll agree that it's worth it. Chelsea Green is a member of the Green Press Initiative (www.greenpressinitiative.org), a nonprofit coalition of publishers, manufacturers, and authors working to protect the world's endangered forests and conserve natural resources. *Walking on Lava* was printed on paper supplied by _____ that contains at least XX% postconsumer recycled fiber.

Library of Congress Cataloging-in-Publication Data
<TK>

Chelsea Green Publishing
85 North Main Street, Suite 120
White River Junction, VT 05001
(802) 295-6300
www.chelseagreen.com

Contents

GARRETT HUPE – Where to? Where from? – *Mount Patterson of the Watupik Range, Alberta, Canada* – Issue 7. Spring 2015

How many ages, how many generations has this great being seen? How many photographs? How many sunsets? How many trees have passed like water beneath its shadow? A mountain just is. Beauty put it there for time to watch.

Uncentring Our Minds

An Introduction to This Book

Uncivilisation, the manifesto that launched the Dark Mountain Project, was written in the autumn of 2008, at a time of global crisis and collapse. A firestorm was blowing through the world's financial system, and for a while it was unclear how much of the world would be left standing when it ended. This book was assembled eight years later, at the end of 2016 — a year in which it was the turn of the West's political systems to feel the force of a storm blowing through history, upending expectations as it passed through.

The timing, in both cases, seems fitting. Dark Mountain was created by two English writers who felt that writing was not doing its job. In a world in which the climate itself was being changed by human activities; in which global ecosystems were dying back before the human advance; and in which the dominant economic and cultural assumptions of the West were clearly beginning to crumble, that little manifesto — twenty pages long, hand-stitched, bound in red paper — asked a simple question: where are the writers, and the artists? Why were the novels, the films, the music, the cultural forms that passed for 'mainstream' in our society still behaving as if it were the twentieth century — or even the nineteenth?

Something else, we thought, was needed. Writers and artists, thinkers and doers, needed to own up to the crises enfolding us, instead of pretending they weren't happening, or that they were just glitches which could be ironed out by technology or politics. An abyss was opening up before us, and the very basis of our civilisation was in question. It was, declared the manifesto, 'time to look down.' If we did, might we be surprised what we saw?

Uncivilisation, like the Modernist manifestos it was inspired by, which had appeared a century before at another time of upheaval, asked these questions to shake people up, to challenge them, to throw cats among pigeons. We wanted writing and art that reflected the dark times we were living in. We wanted writers as prophets, artists who spoke with honest tongues, who might not pretend to have answers but who didn't hide from the questions. We had no idea what this meant, really, or what it would look like, or where to find it, or whether, indeed, it was already out there and we just didn't know about it. The whole thing, like all manifestos, was presumptuous, loud, and arrogant. But it worked.

Nearly ten years on, a tiny initiative launched in the back room of a pub to about forty slightly confused people has become an international network of writers, artists, musicians, thinkers, activists and many others without labels, who believe, as we believed, that honesty about the state of the world is not only necessary but is also a source of energy, and even hope. Once false hopes are shrugged off, once we stare into the darkness that surrounds us, we can see new paths opening up; new forms; new words. The question our manifesto raised – where are all the 'uncivilised' writers and artists – began to be answered almost immediately. They were out there, but they hadn't been able to find each other. What Dark Mountain did – one of the things it did – was to raise a flag around which they could gather.

The results of this gathering have taken many forms. We have run festivals and college courses, worked with national theatres and led mountain expeditions. We have appeared in a strange assortment of media outlets, from radio to TV, from *Resilience.org* to the *New York Times*. We run a website which reaches hundreds of thousands of people each year. We host concerts and lectures, storytellers and performers, poets and magicians.

At the core of all this activity, though, we are doing what we intended to do all along: publishing books. Since 2010, Dark Mountain has published a series of beautifully crafted hardback journals, featuring writing and art from around the world. Each book is packed with contrasting voices and genres: poetry, photography, long essays, flash fiction, paintings, reportage and much else, all brought together by a team of editors many of whom first came to Dark Mountain as unknown voices published in the early books.

All of this has been an attempt to bring together the kind of writing we called for back in 2009:

> It sets out to paint a picture of *homo sapiens* which a being from another world or, better, a being from our own – a blue whale, an albatross, a mountain hare – might recognise as something approaching a truth. It sets out to tug our attention away from ourselves and turn it outwards; to uncentre our minds. It is writing, in short, which puts civilisation – and us – into perspective.

Have we succeeded? That's for others to say. What we can say is that the years which separate the writing of our manifesto from the publishing of this book have seen a widening awareness of the instability of our civilisation and the seriousness of the threats which face it. The kind of arguments which saw us attacked as 'doomers' when we published them in 2010 can be found today on the comment pages of international newspapers. It is becoming harder and harder to deny that what we used to think of as 'progress' is faltering badly. What happens next is the interesting part.

What can writing do about this? What problems can art solve? In one sense, the answers are: nothing, and none. But in another sense, these are the wrong questions. 'All civilisations', we wrote in the manifesto, 'are built on stories.' When the stories fail, we need to know how to tell different ones. We need to have the perspective to understand the failure, and the imagination to offer up new ways of seeing. This is what Dark Mountain set out to do: to play host to voices seeking honest engagement with questions which might be intractable. How else does art get made?

Walking on Lava contains words and images selected from the ten books that the Dark Mountain Project has so far published. It will take you from the mountains of Bolivia to the tribal areas of India and from rural California to the coast of Greenland, in the company of passenger pigeons, squirrels, horses, roe deer and wolves. It will whisk you through time, from medieval Florence to eighteenth century England to prehistoric India to the Younger Dryas. It will introduce you to the history of the future, walk you through the Mahabharata, teach you to think differently about space travel and give you a recipe for pheasant stew.

As we started work on this collection, we wondered how to select from such a diversity of work: what structure could help a newcomer find their way into the territory of Dark Mountain? Then we remembered what we had written at the end of the manifesto. On its last pages, we offered 'eight principles of Uncivilisation': a first attempt to define, point by point, what we stood for and what we wanted to achieve.

The writers, thinkers and artists we have met along this journey have taken us to places that were not marked on our maps, but as we retrace the routes we've explored together, what is striking is how far they have deepened our understanding of each of those principles. So as you follow our footsteps, you will find *Walking on Lava* is organised according to those early categories, with each section of writing introduced by a primal image. This selection is far from a full picture of what we do, but it offers one path across the landscape which Dark Mountain continues to explore. We hope to see you somewhere along the way.

THE EDITORS
JANUARY 2017

UNCIVILISATION
THE DARK MOUNTAIN MANIFESTO

These grand and fatal movements toward death:
        the grandeur of the mass
Makes pity a fool, the tearing pity
For the atoms of the mass, the persons, the victims,
        makes it seem monstrous
To admire the tragic beauty they build.
It is beautiful as a river flowing or a slowly gathering
Glacier on a high mountain rock-face,
Bound to plow down a forest, or as frost in November,
The gold and flaming death-dance for leaves,
Or a girl in the night of her spent maidenhood,
        bleeding and kissing.
I would burn my right hand in a slow fire
To change the future ... I should do foolishly.
        The beauty of modern
Man is not in the persons but in the
Disastrous rhythm, the heavy and mobile masses,
        the dance of the
Dream-led masses down the dark mountain.

Robinson Jeffers, 'Rearmament' (1935)

I

WALKING ON LAVA

The end of the human race will be that it will
eventually die of civilisation.

Ralph Waldo Emerson

Those who witness extreme social collapse at first hand seldom describe any deep revelation about the truths of human existence. What they do mention, if asked, is their surprise at how easy it is to die.

The pattern of ordinary life, in which so much stays the same from one day to the next, disguises the fragility of its fabric. How many of our activities are made possible by the impression of stability that pattern gives? So long as it repeats, or varies steadily enough, we are able to plan for tomorrow as if all the things we rely on and don't think about too carefully will still be there. When the pattern is broken, by civil war or natural disaster or the smaller-scale tragedies that tear at its fabric, many of those activities become impossible or meaningless, while simply meeting needs we once took for granted may occupy much of our lives.

What war correspondents and relief workers report is not only the fragility of the fabric, but the speed with which it can unravel. As we write this, no one can say with certainty where the unravelling of the financial and commercial fabric of our economies will end. Meanwhile, beyond the cities, unchecked industrial exploitation frays the material basis of life in many parts of the world, and pulls at the ecological systems which sustain it.

Precarious as this moment may be, however, an awareness of the fragility of what we call civilisation is nothing new.

'Few men realise,' wrote Joseph Conrad in 1896, 'that their life, the very essence of their character, their capabilities and their audacities, are only the expression of their belief in the safety of their surroundings.' Conrad's writings exposed the civilisation exported by European imperialists to be little more than a comforting illusion, not only in the dark, unconquerable heart of Africa, but in the whited sepulchres of their capital cities. The inhabitants of that civilisation believed 'blindly in the irresistible force of its institutions and its morals, in the power of its police and of its opinion,' but their confidence could be maintained only by the seeming solidity of the crowd of like-minded believers surrounding them. Outside the walls, the wild remained as close to the surface as blood under skin, though the city-dweller was no longer equipped to face it directly.

Bertrand Russell caught this vein in Conrad's worldview, suggesting that the novelist 'thought of civilised and morally tolerable human life as a dangerous walk on a thin crust of barely cooled lava which at any moment might break and let the unwary sink into fiery depths.' What both Russell and Conrad were getting at was a simple fact which any historian could confirm: human civilisation is an intensely fragile construction. It is built on little more than belief: belief in the rightness of its values; belief in the strength of its system of law and order; belief in its currency; above all, perhaps, belief in its future.

Once that belief begins to crumble, the collapse of a civilisation may become unstoppable. That civilisations fall, sooner or later, is as much a law of history as gravity is a law of physics. What remains after the fall is a wild mixture of cultural debris, confused and angry people whose certainties have betrayed them, and those forces which were always there, deeper than the foundations of the city walls: the desire to survive and the desire for meaning.

It is, it seems, our civilisation's turn to experience the inrush of the savage and the unseen; our turn to be brought up short by contact with untamed reality. There is a fall coming. We live in an age in which familiar restraints are being kicked away, and foundations snatched from under us. After a quarter century of complacency, in which we were invited to believe in bubbles that would never burst, prices that would never fall, the end of history, the crude repackaging of the triumphalism of Conrad's Victorian twilight – Hubris has been introduced to Nemesis. Now a familiar human story is being played out. It is the story of an empire corroding from within. It is the story of a people who believed, for a long time, that their actions did not have consequences. It is the story of how that people will cope with the crumbling of their own myth. It is our story.

This time, the crumbling empire is the unassailable global economy, and the brave new world of consumer democracy being forged worldwide in its name. Upon the indestructibility of this edifice we have pinned the hopes of this latest phase of our civilisation. Now, its failure and fallibility exposed, the world's elites are scrabbling frantically to buoy up an economic machine which, for decades, they told us needed little restraint, for restraint would be its undoing. Uncountable sums of money are being funnelled upwards in order to prevent an uncontrolled explosion. The machine is stuttering and the engineers are in panic. They are wondering if perhaps they do not understand it as well as they imagined. They are wondering whether they are controlling it at all or whether, perhaps, it is controlling them.

Increasingly, people are restless. The engineers group themselves into competing teams, but neither side seems to know what to do, and neither seems much different from the other. Around the world, discontent can be heard. The extremists are grinding their knives and moving in as the machine's coughing and stuttering exposes the inadequacies of the political oligarchies who claimed to have everything in hand. Old gods are rearing their heads, and old answers: revolution, war, ethnic strife. Politics as we have known it totters, like the machine it was built to sustain. In its place could easily arise something more elemental, with a dark heart.

As the financial wizards lose their powers of levitation, as the politicians and economists struggle to conjure new explanations, it starts to dawn on us that behind the curtain, at the heart of the Emerald City, sits not the benign and omnipotent invisible hand we had been promised, but something else entirely. Something responsible for what Marx, writing not so long before Conrad, cast as the 'everlasting uncertainty and anguish' of the 'bourgeois epoch'; a time in which 'all that is solid melts into air, all that is holy is profaned.' Draw back the curtain, follow the tireless motion of cogs and wheels back to its source, and you will find the engine driving our civilisation: the myth of progress.

The myth of progress is to us what the myth of god-given warrior prowess was to the Romans, or the myth of eternal salvation was to the conquistadors: without it, our efforts cannot be sustained. Onto the rootstock of Western Christianity, the Enlightenment at its most optimistic grafted a vision of an Earthly paradise, towards which human effort guided by calculative reason could take us. Following this guidance, each generation will live a better life than the life of those that went before it. History becomes an escalator, and the only way is up. On the top floor is human perfection. It is important that this should remain just out of reach in order to sustain the sensation of motion.

Recent history, however, has given this mechanism something of a battering. The past century too often threatened a descent into hell, rather than the promised heaven on Earth. Even within the prosperous and liberal societies of the West progress has, in many ways, failed to deliver the goods. Today's generation are demonstrably less content, and consequently less optimistic, than those that went before. They work longer hours, with less security, and less chance of leaving behind the social background into which they were born. They fear crime, social breakdown, overdevelopment, environmental collapse. They do not believe that the future will be better than the past. Individually, they are less constrained by class and convention than their parents or grandparents, but more constrained by law, surveillance, state proscription and personal debt. Their physical health is better, their mental health more fragile. Nobody knows what is coming. Nobody wants to look.

Most significantly of all, there is an underlying darkness at the root of everything we have built. Outside the cities, beyond the blurring edges of our civilisation, at the mercy of the machine but not under its control, lies something that neither Marx nor Conrad, Caesar nor Hume, Thatcher nor Lenin ever really understood. Something that Western civilisation — which has set the terms for global civilisation — was never capable of understanding, because to understand it would be to undermine, fatally, the myth of that civilisation. Something upon which that thin crust of lava is balanced; which feeds the machine and all the people who run it, and which they have all trained themselves not to see.

II

THE SEVERED HAND

Then what is the answer? Not to be deluded by dreams.
To know that great civilisations have broken down into violence,
  and their tyrants come, many times before.
When open violence appears, to avoid it with honor or choose
  the least ugly faction; these evils are essential.
To keep one's own integrity, be merciful and uncorrupted
  and not wish for evil; and not be duped
By dreams of universal justice or happiness. These dreams will
  not be fulfilled.
To know this, and know that however ugly the parts appear
  the whole remains beautiful. A severed hand
Is an ugly thing and man dissevered from the earth and stars
  and his history ... for contemplation or in fact ...
Often appears atrociously ugly. Integrity is wholeness,
  the greatest beauty is
Organic wholeness, the wholeness of life and things, the divine beauty
  of the universe. Love that, not man
Apart from that, or else you will share man's pitiful confusions,
  or drown in despair when his days darken.

Robinson Jeffers, 'The Answer'

The myth of progress is founded on the myth of nature. The first tells us that we are destined for greatness; the second tells us that greatness is cost-free. Each is intimately bound up with the other. Both tell us that we are apart from the world; that we began grunting in the primeval swamps, as a humble part of something called 'nature', which we have now triumphantly subdued. The very fact that we have a word for 'nature' is evidence that we do not regard ourselves as part of it. Indeed, our separation from it is a myth integral to the triumph of our civilisation. We are, we tell ourselves, the only species ever to have attacked nature and won. In this, our unique glory is contained.

Outside the citadels of self-congratulation, lone voices have cried out against this infantile version of the human story for centuries, but it is only in the last few decades that its inaccuracy has become laughably apparent. We are the first generations to grow up surrounded by evidence that our attempt to separate ourselves from 'nature' has been a grim failure, proof not of our genius but our hubris. The attempt to sever the hand from the body has endangered the 'progress' we hold so dear, and it has endangered much of 'nature' too. The resulting upheaval underlies the crisis we now face.

We imagined ourselves isolated from the source of our existence. The fallout from this imaginative error is all around us: a quarter of the world's mammals are threatened with imminent extinction; an acre and a half of rainforest is felled every second; 75% of the world's fish stocks are on the verge of collapse; humanity consumes 25% more of the world's natural 'products' than the Earth can replace – a figure predicted to rise to 80% by mid-century. Even through the deadening lens of statistics, we can glimpse the violence to which our myths have driven us.

And over it all looms runaway climate change. Climate change, which threatens to render all human projects irrelevant; which presents us with detailed evidence of our lack of understanding of the world we inhabit while, at the same time, demonstrating that we are still entirely reliant upon it. Climate change, which highlights in painful colour the head-on crash between civilisation and 'nature'; which makes plain, more effectively than any carefully constructed argument or optimistically defiant protest, how the machine's need for permanent growth will require us to destroy ourselves in its name. Climate change, which brings home at last our ultimate powerlessness.

These are the facts, or some of them. Yet facts never tell the whole story. ('Facts', Conrad wrote, in Lord Jim, 'as if facts could prove anything.') The facts of environmental crisis we hear so much about often conceal as much as they expose. We hear daily about the impacts of our activities on 'the environment' (like 'nature', this is an expression which distances us from the reality of our situation). Daily we hear, too, of the many 'solutions' to these problems: solutions which usually involve the necessity of urgent political agreement and a judicious application of human technological genius. Things may be changing, runs the narrative, but there is nothing we cannot deal with here, folks. We perhaps need to move faster, more urgently. Certainly we need to accelerate the pace of research and development. We accept that we must become more 'sustainable'. But everything will be fine. There will still be growth, there will still be progress: these things will continue, because they have to continue, so they cannot do anything but continue. There is nothing to see here. Everything will be fine.

❧

We do not believe that everything will be fine. We are not even sure, based on current definitions of progress and improvement, that we want it to be. Of all humanity's delusions of difference, of its separation from and superiority to the living world which surrounds it, one distinction holds up better than most: we may well be the first species capable of effectively eliminating life on Earth. This is a hypothesis we seem intent on putting to the test. We are already responsible for denuding the world of much of its richness, magnificence, beauty, colour and magic, and we show no sign of slowing down. For a very long time, we imagined that 'nature' was something that happened elsewhere. The damage we did to it might be regrettable, but needed to be weighed against the benefits here and now. And in the worst case scenario, there would always be some kind of Plan B. Perhaps we would make for the moon, where we could survive in lunar colonies under giant bubbles as we planned our expansion across the galaxy.

But there is no Plan B and the bubble, it turns out, is where we have been living all the while. The bubble is that delusion of isolation under which we have laboured for so long. The bubble has cut us off from life on the only planet we have, or are ever likely to have. The bubble is civilisation.

Consider the structures on which that bubble has been built. Its foundations are geological: coal, oil, gas – millions upon millions of years of ancient sunlight, dragged from the depths of the planet and burned with abandon. On this base, the structure stands. Move upwards, and you pass through a jumble of supporting horrors: battery chicken sheds; industrial abattoirs; burning forests; beam-trawled ocean floors; dynamited reefs; hollowed-out mountains; wasted soil. Finally, on top of all these unseen layers, you reach the well-tended surface where you and I stand: unaware, or uninterested, in what goes on beneath us; demanding that the authorities keep us in the manner to which we have been accustomed; occasionally feeling twinges of guilt that lead us to buy organic chickens or locally-produced lettuces; yet for the most part glutted, but not sated, on the fruits of the horrors on which our lifestyles depend.

We are the first generations born into a new and unprecedented age – the age of ecocide. To name it thus is not to presume the outcome, but simply to describe a process which is underway. The ground, the sea, the air, the elemental backdrops to our existence — all these our economics has taken for granted, to be used as a bottomless tip, endlessly able to dilute and disperse the tailings of our extraction, production, consumption. The sheer scale of the sky or the weight of a swollen river makes it hard to imagine that creatures as flimsy as you and I could do that much damage. Philip Larkin gave voice to this attitude, and the creeping, worrying end of it in his poem 'Going, Going':

> Things are tougher than we are, just
> As earth will always respond
> However we mess it about;
> Chuck filth in the sea, if you must:
> The tides will be clean beyond.
> — But what do I feel now? Doubt?

Nearly forty years on from Larkin's words, doubt is what all of us seem to feel, all of the time. Too much filth has been chucked in the sea and into the soil and into the atmosphere to make any other feeling sensible. The doubt, and the facts, have paved the way for a worldwide movement of environmental politics, which aimed, at least in its early, raw form, to challenge the myths of development and progress head-on. But time has not been kind to the greens. Today's environmentalists are more likely to be found at corporate conferences hymning the virtues of 'sustainability' and 'ethical consumption' than doing anything as naive as questioning the intrinsic values of civilisation. Capitalism has absorbed the greens, as it absorbs so many challenges to its ascendancy. A radical challenge to the human machine has been transformed into yet another opportunity for shopping.

'Denial' is a hot word, heavy with connotations. When it is used to brand the remaining rump of climate change sceptics, they object noisily to the association with those who would rewrite the history of the Holocaust. Yet the focus on this dwindling group may serve as a distraction from a far larger form of denial, in its psychoanalytic sense. Freud wrote of the inability of people to hear things which did not fit with the way they saw themselves and the world. We put ourselves through all kinds of inner contortions, rather than look plainly at those things which challenge our fundamental understanding of the world.

Today, humanity is up to its neck in denial about what it has built, what it has become — and what it is in for. Ecological and economic collapse unfold before us and, if we acknowledge them at all, we act as if this were a temporary problem, a technical glitch. Centuries of hubris block our ears like wax plugs; we cannot hear the message which reality is screaming at us.

For all our doubts and discontents, we are still wired to an idea of history in which the future will be an upgraded version of the present. The assumption remains that things must continue in their current direction: the sense of crisis only smudges the meaning of that 'must'. No longer a natural inevitability, it becomes an urgent necessity: we must find a way to go on having supermarkets and superhighways. We cannot contemplate the alternative.

And so we find ourselves, all of us together, poised trembling on the edge of a change so massive that we have no way of gauging it. None of us knows where to look, but all of us know not to look down. Secretly, we all think we are doomed: even the politicians think this; even the environmentalists. Some of us deal with it by going shopping. Some deal with it by hoping it is true. Some give up in despair. Some work frantically to try and fend off the coming storm.

Our question is: what would happen if we looked down? Would it be as bad as we imagine? What might we see? Could it even be good for us?

We believe it is time to look down.

III

UNCIVILISATION

*Without mystery, without curiosity and without the form imposed by
a partial answer, there can be no stories—only confessions, commu-
niqués, memories and fragments of autobiographical fantasy which
for the moment pass as novels.*

John Berger, 'A Story for Aesop'

If we are indeed teetering on the edge of a massive change in how we live,
in how human society itself is constructed, and in how we relate to the rest
of the world, then we were led to this point by the stories we have told ourselves — above all, by the story of civilisation.

This story has many variants, religious and secular, scientific, economic
and mystic. But all tell of humanity's original transcendence of its animal
beginnings, our growing mastery over a 'nature' to which we no longer
belong, and the glorious future of plenty and prosperity which will follow
when this mastery is complete. It is the story of human centrality, of a species destined to be lord of all it surveys, unconfined by the limits that apply
to other, lesser creatures.

What makes this story so dangerous is that, for the most part, we have
forgotten that it is a story. It has been told so many times by those who see
themselves as rationalists, even scientists; heirs to the Enlightenment's legacy
— a legacy which includes the denial of the role of stories in making the world.

Humans have always lived by stories, and those with skill in telling
them have been treated with respect and, often, a certain wariness. Beyond
the limits of reason, reality remains mysterious, as incapable of being
approached directly as a hunter's quarry. With stories, with art, with sym-
bols and layers of meaning, we stalk those elusive aspects of reality that go
undreamed of in our philosophy. The storyteller weaves the mysterious into
the fabric of life, lacing it with the comic, the tragic, the obscene, making
safe paths through dangerous territory.

Yet as the myth of civilisation deepened its grip on our thinking, borrowing the guise of science and reason, we began to deny the role of stories, to
dismiss their power as something primitive, childish, outgrown. The old

tales by which generations had made sense of life's subtleties and strange-
nesses were bowdlerised and packed off to the nursery. Religion, that bag of
myths and mysteries, birthplace of the theatre, was straightened out into a
framework of universal laws and moral account-keeping. The dream visions
of the Middle Ages became the nonsense stories of Victorian childhood. In
the age of the novel, stories were no longer the way to approach the deep
truths of the world, so much as a way to pass time on a train journey. It is
hard, today, to imagine that the word of a poet was once feared by a king.

Yet for all this, our world is still shaped by stories. Through television,
film, novels and video games, we may be more thoroughly bombarded with
narrative material than any people that ever lived. What is peculiar, however, is the carelessness with which these stories are channelled at us – as
entertainment, a distraction from daily life, something to hold our attention
to the other side of the ad break. There is little sense that these things make
up the equipment by which we navigate reality. On the other hand, there are
the serious stories told by economists, politicians, geneticists and corporate
leaders. These are not presented as stories at all, but as direct accounts of
how the world is. Choose between competing versions, then fight with those
who chose differently. The ensuing conflicts play out on early morning
radio, in afternoon debates and late night television pundit wars. And yet,
for all the noise, what is striking is how much the opposing sides agree on:
all their stories are only variants of the larger story of human centrality, of
our ever-expanding control over 'nature', our right to perpetual economic
growth, our ability to transcend all limits.

So we find ourselves, our ways of telling unbalanced, trapped inside a
runaway narrative, headed for the worst kind of encounter with reality. In
such a moment, writers, artists, poets and storytellers of all kinds have a
critical role to play. Creativity remains the most uncontrollable of human
forces: without it, the project of civilisation is inconceivable, yet no part
of life remains so untamed and undomesticated. Words and images can
change minds, hearts, even the course of history. Their makers shape the
stories people carry through their lives, unearth old ones and breathe them
back to life, add new twists, point to unexpected endings. It is time to pick
up the threads and make the stories new, as they must always be made new,
starting from where we are.

Mainstream art in the West has long been about shock; about busting
taboos, about Getting Noticed. This has gone on for so long that it has
become common to assert that in these ironic, exhausted, post-everything
times, there are no taboos left to bust. But there is one.

The last taboo is the myth of civilisation. It is built upon the stories we have constructed about our genius, our indestructibility, our manifest destiny as a chosen species. It is where our vision and our self-belief intertwine with our reckless refusal to face the reality of our position on this Earth. It has led the human race to achieve what it has achieved; and has led the planet into the age of ecocide. The two are intimately linked. We believe they must be decoupled if anything is to remain.

We believe that artists – which is to us the most welcoming of words, taking under its wing writers of all kinds, painters, musicians, sculptors, poets, designers, creators, makers of things, dreamers of dreams – have a responsibility to begin the process of decoupling. We believe that, in the age of ecocide, the last taboo must be broken – and that only artists can do it.

Ecocide demands a response. That response is too important to be left to politicians, economists, conceptual thinkers, number crunchers; too all-pervasive to be left to activists or campaigners. Artists are needed. So far, though, the artistic response has been muted. In between traditional nature poetry and agitprop, what is there? Where are the poems that have adjusted their scope to the scale of this challenge? Where are the novels that probe beyond the country house or the city centre? What new form of writing has emerged to challenge civilisation itself? What gallery mounts an exhibition equal to this challenge? Which musician has discovered the secret chord?

If the answers to these questions have been scarce up to now, it is perhaps both because the depth of collective denial is so great, and because the challenge is so very daunting. We are daunted by it, ourselves. But we believe it needs to be risen to. We believe that art must look over the edge, face the world that is coming with a steady eye, and rise to the challenge of ecocide with a challenge of its own: an artistic response to the crumbling of the empires of the mind.

This response we call Uncivilised art, and we are interested in one branch of it in particular: Uncivilised writing. Uncivilised writing is writing which attempts to stand outside the human bubble and see us as we are: highly evolved apes with an array of talents and abilities which we are unleashing without sufficient thought, control, compassion or intelligence. Apes who have constructed a sophisticated myth of their own importance with which to sustain their civilising project. Apes whose project has been to tame, to control, to subdue or to destroy — to civilise the forests, the deserts, the wild lands and the seas, to impose bonds on the minds of their own in order that they might feel nothing when they exploit or destroy their fellow creatures.

Against the civilising project, which has become the progenitor of ecocide, Uncivilised writing offers not a non-human perspective—we remain human and, even now, are not quite ashamed – but a perspective which sees us as one strand of a web rather than as the first palanquin in a glorious procession. It offers an unblinking look at the forces among which we find ourselves.

It sets out to paint a picture of *homo sapiens* which a being from another world or, better, a being from our own – a blue whale, an albatross, a mountain hare – might recognise as something approaching a truth. It sets out to tug our attention away from ourselves and turn it outwards; to uncentre our minds. It is writing, in short, which puts civilisation – and us – into perspective. Writing that comes not, as most writing still does, from the self-absorbed and self-congratulatory metropolitan centres of civilisation but from somewhere on its wilder fringes. Somewhere woody and weedy and largely avoided, from where insistent, uncomfortable truths about ourselves drift in; truths which we're not keen on hearing. Writing which unflinchingly stares us down, however uncomfortable this may prove.

It might perhaps be just as useful to explain what Uncivilised writing is not. It is not environmental writing, for there is much of that about already, and most of it fails to jump the barrier which marks the limit of our collective human ego; much of it, indeed, ends up shoring-up that ego, and helping us to persist in our civilisational delusions. It is not nature writing, for there is no such thing as nature as distinct from people, and to suggest otherwise is to perpetuate the attitude which has brought us here. And it is not political writing, with which the world is already flooded, for politics is a human confection, complicit in ecocide and decaying from within.

Uncivilised writing is more rooted than any of these. Above all, it is determined to shift our worldview, not to feed into it. It is writing for outsiders. If you want to be loved, it might be best not to get involved, for the world, at least for a time, will resolutely refuse to listen.

A salutary example of this last point can be found in the fate of one of the twentieth century's most significant yet most neglected poets. Robinson Jeffers was writing Uncivilised verse seventy years before this manifesto was thought of, though he did not call it that. In his early poetic career, Jeffers was a star: he appeared on the cover of Time magazine, read his poems in the US Library of Congress and was respected for the alternative he offered to the Modernist juggernaut. Today his work is left out of anthologies, his name is barely known and his politics are regarded with suspicion. Read Jeffers' later work and you will see why. His crime was to deliberately puncture humanity's sense of self-importance. His punishment was to be sent into a lonely literary exile from which, forty years after his death, he has still not been allowed to return.

But Jeffers knew what he was in for. He knew that nobody, in an age of 'consumer choice', wanted to be told by this stone-faced prophet of the California cliffs that 'it is good for man ... To know that his needs and nature are no more changed in fact in ten thousand years than the beaks of eagles.' He knew that no comfortable liberal wanted to hear his angry warning, issued at the height of the Second World War: 'Keep clear of the dupes that talk democracy / And the dogs that talk revolution / Drunk with talk, liars and believers ... / Long live freedom, and damn the ideologies.' His vision of a world in which humanity was doomed to destroy its surroundings and eventually itself ('I would burn my right hand in a slow fire / To change the future ... I should do foolishly') was furiously rejected in the rising age of consumer democracy which he also predicted ('Be happy, adjust your economics to the new abundance...')

Jeffers, as his poetry developed, developed a philosophy too. He called it 'inhumanism.' It was, he wrote:

> a shifting of emphasis and significance from man to notman; the rejection of human solipsism and recognition of the trans-human magnificence...This manner of thought and feeling is neither misanthropic nor pessimist ... It offers a reasonable detachment as rule of conduct, instead of love, hate and envy... it provides magnificence for the religious instinct, and satisfies our need to admire greatness and rejoice in beauty.

The shifting of emphasis from man to notman: this is the aim of Uncivilised writing. To 'unhumanise our views a little, and become confident / As the rock and ocean that we were made from.' This is not a rejection of our humanity – it is an affirmation of the wonder of what it means to be truly human. It is to accept the world for what it is and to make our home here, rather than dreaming of relocating to the stars, or existing in a Man-forged bubble and pretending to ourselves that there is nothing outside it to which we have any connection at all.

This, then, is the literary challenge of our age. So far, few have taken it up. The signs of the times flash out in urgent neon, but our literary lions have better things to read. Their art remains stuck in its own civilised bubble. The idea of civilisation is entangled, right down to its semantic roots, with city-dwelling, and this provokes a thought: if our writers seem unable to find new stories which might lead us through the times ahead, is this not a function of their metropolitan mentality? The big names of contemporary literature are equally at home in the fashionable quarters of London or New York, and their writing reflects the prejudices of the placeless, transnational elite to which they belong.

The converse also applies. Those voices which tell other stories tend to be rooted in a sense of place. Think of John Berger's novels and essays from the Haute Savoie, or the depths explored by Alan Garner within a day's walk of his birthplace in Cheshire. Think of Wendell Berry or WS Merwin, Mary Oliver or Cormac McCarthy. Those whose writings approach the shores of the Uncivilised are those who know their place, in the physical sense, and who remain wary of the siren cries of metrovincial fashion and civilised excitement.

If we name particular writers whose work embodies what we are arguing for, the aim is not to place them more prominently on the existing map of literary reputations. Rather, as Geoff Dyer has said of Berger, to take their work seriously is to redraw the maps altogether – not only the map of literary reputations, but those by which we navigate all areas of life.

Even here, we go carefully, for cartography itself is not a neutral activity. The drawing of maps is full of colonial echoes. The civilised eye seeks to view the world from above, as something we can stand over and survey. The Uncivilised writer knows the world is, rather, something we are enmeshed in — a patchwork and a framework of places, experiences, sights, smells, sounds. Maps can lead, but can also mislead. Our maps must be the kind sketched in the dust with a stick, washed away by the next rain. They can be read only by those who ask to see them, and they cannot be bought.

This, then, is Uncivilised writing. Human, inhuman, stoic and entirely natural. Humble, questioning, suspicious of the big idea and the easy answer. Walking the boundaries and reopening old conversations. Apart but engaged, its practitioners always willing to get their hands dirty; aware, in fact, that dirt is essential; that keyboards should be tapped by those with soil under their fingernails and wilderness in their heads.

We tried ruling the world; we tried acting as God's steward, then we tried ushering in the human revolution, the age of reason and isolation. We failed in all of it, and our failure destroyed more than we were even aware of. The time for civilisation is past. Uncivilisation, which knows its flaws because it has participated in them; which sees unflinchingly and bites down hard as it records – this is the project we must embark on now. This is the challenge for writing – for art – to meet. This is what we are here for.

IV

TO THE FOOTHILLS!

One impulse from a vernal wood
May teach you more of man,
Of moral evil and of good,
Than all the sages can.

William Wordsworth, 'The Tables Turned'

A movement needs a beginning. An expedition needs a base camp. A project needs a headquarters. Uncivilisation is our project, and the promotion of Uncivilised writing – and art needs a base. We present this manifesto not simply because we have something to say —who doesn't? —but because we have something to do. We hope this pamphlet has created a spark. If so, we have a responsibility to fan the flames. This is what we intend to do. But we can't do it alone.

This is a moment to ask deep questions and to ask them urgently. All around us, shifts are under way which suggest that our whole way of living is already passing into history. It is time to look for new paths and new stories, ones that can lead us through the end of the world as we know it and out the other side. We suspect that by questioning the foundations of civilisation, the myth of human centrality, our imagined isolation, we may find the beginning of such paths.

If we are right, it will be necessary to go literally beyond the Pale. Outside the stockades we have built — the city walls, the original marker in stone or wood that first separated 'man' from 'nature'. Beyond the gates, out into the wilderness, is where we are headed. And there we shall make for the higher ground for, as Jeffers wrote, 'when the cities lie at the monster's feet / There are left the mountains.' We shall make the pilgrimage to the poet's Dark Mountain, to the great, immovable, inhuman heights which were here before us and will be here after, and from their slopes we shall look back upon the pinprick lights of the distant cities and gain perspective on who we are and what we have become.

This is the Dark Mountain project. It starts here.

Where will it end? Nobody knows. Where will it lead? We are not sure. Its first incarnation, launched alongside this manifesto, is a website, which points the way to the ranges. It will contain thoughts, scribblings, jottings, ideas; it will work up the project of Uncivilisation, and invite all comers to join the discussion.

Then it will become a physical object, because virtual reality is, ultimately, no reality at all. It will become a journal, of paper, card, paint and print; of ideas, thoughts, observations, mumblings; new stories which will help to define the project – the school, the movement – of Uncivilised writing. It will collect the words and the images of those who consider themselves Uncivilised and have something to say about it; who want to help us attack the citadels. It will be a thing of beauty for the eye and for the heart and for the mind, for we are unfashionable enough to believe that beauty – like truth – not only exists, but still matters.

Beyond that... all is currently hidden from view. It is a long way across the plains, and things become obscured by distance. There are great white spaces on this map still. The civilised would fill them in; we are not so sure we want to. But we cannot resist exploring them, navigating by rumours and by the stars. We don't know quite what we will find. We are slightly nervous. But we will not turn back, for we believe that something enormous may be out there, waiting to meet us.

Uncivilisation, like civilisation, is not something that can be created alone. Climbing the Dark Mountain cannot be a solitary exercise. We need bearers, sherpas, guides, fellow adventurers. We need to rope ourselves together for safety. At present, our form is loose and nebulous. It will firm itself up as we climb. Like the best writing, we need to be shaped by the ground beneath our feet, and what we become will be shaped, at least in part, by what we find on our journey.

If you would like to climb at least some of the way with us, we would like to hear from you. We feel sure there are others out there who would relish joining us on this expedition.

Come. Join us. We leave at dawn.

THE EIGHT PRINCIPLES OF UNCIVILISATION

'We must unhumanise our views a little, and become confident
As the rock and ocean that we were made from.'

1. We live in a time of social, economic and ecological unravelling. All around us are signs that our whole way of living is already passing into history. We will face this reality honestly and learn how to live with it.

2. We reject the faith which holds that the converging crises of our times can be reduced to a set of 'problems' in need of technological or political 'solutions'.

3. We believe that the roots of these crises lie in the stories we have been telling ourselves. We intend to challenge the stories which underpin our civilisation: the myth of progress, the myth of human centrality, and the myth of our separation from 'nature'. These myths are more dangerous for the fact that we have forgotten they are myths.

4. We will reassert the role of storytelling as more than mere entertainment. It is through stories that we weave reality.

5. Humans are not the point and purpose of the planet. Our art will begin with the attempt to step outside the human bubble. By careful attention, we will reengage with the non-human world.

6. We will celebrate writing and art which is grounded in a sense of place and of time. Our literature has been dominated for too long by those who inhabit the cosmopolitan citadels.

7. We will not lose ourselves in the elaboration of theories or ideologies. Our words will be elemental. We write with dirt under our fingernails.

8. The end of the world as we know it is not the end of the world full stop. Together, we will find the hope beyond hope, the paths which lead to the unknown world ahead of us.

PAUL KINGSNORTH AND DOUGALD HINE
OXFORD, ENGLAND 2009

ONE

We live in a time of social, economic and ecological unravelling. All around us are signs that our whole way of living is already passing into history. We will face this reality honestly and learn how to live with it.

RON HAGG – Adobe Farmhouse – *New Mexico, USA* – *Issue 7. Spring 2015*

This is an abandoned adobe farmhouse in Arroyo Hondo, New Mexico. It is sad to see a family's dreams abandoned where once there was hope. Today in the United States we are run by a corporate state – small family farms are disappearing at an alarming rate, only to taken over by huge factory farms. In the photo-graph I try to capture not only a devastating case of a single family no longer farming, but also a reflection of the state of affairs in our society. Before the arrival of the Europeans in 1700 this area was populated by members of the Taos pueblo.

Early Knowledge

ANITA SULLIVAN

Issue 5, Spring 2014

In paradise Adam stubbed his toe on a rock.
It bled and bled, until, at the snake's suggestion
he wrapped it in a poultice of dampened leaves
from the Forbidden Tree, thus gaining
through osmosis
a wholly different kind of knowledge.

After Adam's foot healed he
could walk across hot coals and not be burned.

Everywhere the family went outside the walls
the sun was local,
already had a name like Hank
or That-Which-Makes-Rocks-Steam.

The moon too, and most of
the animals had different names
than the ones he first gave them.

But with his new knowledge
Adam could go deeper.
Earth, Air, Fire, Water — he broke them down
into smaller and smaller parts:
            electron   proton   Higgs boson
neutrino   quark ...

In his final decades
Adam went over to naming horses
thus assuring an endless supply for future races.
Eve would find him
            evenings
under the village bougainvillea

reciting into his beer

Black Caviar   Asteroid   Crucifix
Awesome Feather   Bustin Stones   Queen's Logic
Wounded Knee ...

Eve took a lover from one of the Nephilim.

Loss Soup

NICK HUNT

Issue 1, Summer 2010

FIGURE 1A: THE DINING HALL. Located, it seems, in an abandoned subway tunnel, panelled incongruously in teak, mahogany and other unsustainable hardwoods. Insufficiently lit by recessed lights that give the room an atmosphere of twilight. Walls cluttered dustily with half-completed objects, broken bits of statuary that appear familiar at first glance, and at second glance unrecognisable. Things that make you say to yourself, 'I'll have a closer look at that later,' but, of course, you never do.

FIGURE 1B: THE DINING TABLE. It stretches the full length of the hall, and appears to be constructed from railway sleepers, or planks from some old galleon. It must weigh many tonnes. Glancing beneath, you see it is supported by a forest of legs of many different shapes and sizes, cannibalised from tables, chairs, pedestals, crutches, walking sticks. Laid out upon the bare expanse of wood are two rows of dusty glasses, two rows of earthenware bowls, and some wooden spoons.

FIGURE 1C: THE DINERS. At first you assume there are scores of them, but later adjust your estimate to only a few dozen. Calculating numbers is surprisingly tricky, due to the dimness of the light and the peculiar amorphousness of facial features. Various races are represented, there's an equal ratio of women to men, but around this table they all appear generic. It's not helped by the fact they keep changing position without you noticing them move. You turn away from the man to your left, a Slavic gentleman with impressive moustaches, and when you turn back it's an old Asian lady with spectacles like the lenses from antique telescopes. But it's hard to be sure. Your concentration keeps slipping. Perhaps this is still the same person, with a different facial expression.

FIGURE 2: THE EGG-TIMER (A). It stands at the furthest end of the table, about the height of a grandfather clock, a truly impressive object. A baroque monstrosity of piped and fluted metal, like something from the palace of the Tsars. The dirty golden sand hisses audibly from the top chamber to the bottom, and an ingenious pivoting mechanism allows the whole thing to be rotated when the bottom chamber is full. This task, you imagine, will be performed by the diners sitting on either side, who are watching the sand's flow closely. But the top chamber isn't empty yet.

FIGURE 3A: THE SOUP TUREEN. It is wheeled in on a serving trolley, and lifted onto the table by three waiters. Its arrival elicits little excitement from the assembled diners, though you, a first-timer, are awed by its size. 'Could fit a whole lot of soup in there,' you scribble on the first page of your notebook. But the tureen, as far as you see, has yet to be filled.

FIGURE 3B: THE LADLE. It's a big one.

FIGURE 4: THE OBSERVER. This is you. You still can't quite believe you've been chosen to attend the fabled Dinner of Loss, but here you sit, notebook on table, wooden spoon in hand. A poorly accredited freelance journalist with a vague interest in 'disappearing things' – you've written articles on language extinction, vanishing glaciers, memory loss – you received the invitation three days ago, and cancelled all previous engagements. You've come across mention of the Dinner of Loss in the course of your researches, of course, but were doubtful if the rumours were true. As far as you know, one lucky observer is invited to attend every year, but you can't imagine how the organisers came to choose you.

You came here in an ordinary taxi, though half expecting to be blindfolded and spun around for disorientation. You entered through an ordinary door, following the instructions. You descended several flights of stairs, walked down a mothball-smelling corridor, entered the long dining hall, and found your place-name waiting.

You've been here about forty-five minutes. The dinner is due to begin.

FIGURE 5: THE GONG. It gongs. A silence settles around the table.

FIGURE 6: THE FIRST INTONATIONS. Delivered by one diner after another, passing around the table in turn, at a steady metronomic pace, in an anti-clockwise direction. Running, as far as you can note, as follows:

'The auroch. The Barbary lion. The Japanese wolf. The giant short-faced bear. The upland moa. The American bison. The broad-faced potoroo. The American lion. The elephant bird. The Caucasian wisent. The cave bear. The passenger *pigeon. The Nendo tube-nosed fruit bat. The Darling Downs hopping mouse. The dwarf elephant. The Syrian wild ass. The Rabb's fringe-limbed tree frog. The St Lucy giant rice rat ...'*

You scribble as fast as your biro can go, but the separately spoken intonations dissolve into a quiet cacophony, murmuring like a disturbed sea, with little rhyme or rhythm. They don't appear to follow any order, whether categorical or chronological. Your writing degrades into improvised shorthand you're not even sure you'll be able to read.

'The ground sloth. The pig-footed bandicoot. The Balearic shrew. The Ilin Island cloudrunner. The Arabian gazelle. The Schomburgk's deer. The sea mink. The Javan tiger. The tarpan. The great auk. The Alaotra grebe. The Bermuda night heron. The laughing owl. The bluebuck. The quagga. The western black rhinoceros. The Sturdee's pipistrelle. The thylacine. The turquoise-throated puffleg ...'

At last the intonations stop. Page after page of your notebook is covered in increasingly frenetic scrawls. You think perhaps an hour has passed, but since they removed your watch at the door you have no way of knowing. The only indicator of time is the giant egg-timer down the table, the snakey sand still hissing inside, though the top chamber still isn't empty. Your writing hand throbs, and you're glad of the few minutes' interregnum in which each diner finds their glass has been filled with wine at some point during the proceedings. Following the lead of the other diners, you raise your glass into the air, casting wobbling wine-shadows over the wood.

'Lost animals,' a voice concludes. And as the glasses chime together, the trio of waiters re-enters the hall bearing a steaming vat.

FIGURE 7: LOSS SOUP (A). The waiters approach the soup tureen. You rise from your chair to get a better look, thrilled to be witness to the fabled soup itself, and a slight tut-tut of disapproval issues from the diners beside you. You disregard this. You're a journalist. You can't help but elicit disapproval at times. You lean across the table, on tiptoes, to get closer to the action.

Actually there isn't much to see. The waiters remove the tureen's heavy lid and upend the steaming vat. You strain to get a good look at the soup as it sloppily cascades into the tureen, but all you can make out is a viscous gruel, thickened occasionally with matter you can't from this distance identify, a greasy sludge of no definable colour. Although the vat is of no small proportions, you guess the soup that has been poured must cover only an inch or two at the base of the vast tureen. When the gush comes to an end, the waiters shake the last drops out, replace the cumbersome china lid, bow to no-one in particular, and retire.

FIGURE 8: THE SECOND INTONATIONS. Before you are even resettled in your seat, the next round has begun.

'Geeze. Nagumi. Kw'adza. Eyak. Esselen. Island Chumash. Hittite. Eel River Athabaskan. Lycian. Kalkatungic. Moabite. Coptic. Oti. Karipuna. Totoro. Ancient Nubian. Yahuna. Wasu. Old Prussian. Old Tatar. Modern Gutnish. Skepi Creole Dutch ... '

You begin to feel a little light-headed. Your biro loses track. You are forced to resort to abbreviations you despair of ever deciphering. But still, you must attempt to keep pace with the murmuring litany of names, must try to record as many as you can, for they are fast disappearing.

The air itself seems to draw them in. They have no body, no substance. The sounds are like vapour, amorphous, removed from reality.

'Akkala Sámi. Old Church Slavonic. Bo. Kseireins. Scythian. Cuman. Pictish. Karnic. Etruscan. Wagaya-Warluwaric. Edomite. Tangut. Ammonite. Minaean. Phoenician. Ugaritic. Basque-Icelandic pidgin ... '

'Lost languages,' the soft voice says, dropping at last a tangible sound – if there can exist such a thing – into a silence you hadn't been made aware of. Glasses clink. You have missed the toast. You are still trying to scribble the last names before the sounds go out of your head. But it's no good, you can't remember.

FIGURE 9: LOSS SOUP (B). Again the waiters bring the vat, and you get to your feet to see the gruel slide like an oil slick into the tureen, billowing up clouds of steam. It gives a thin, faintly saline smell. The lid is replaced. The table settles down. The sand inside the egg-timer whispers to itself in the corner.

FIGURE 10: THE THIRD INTONATIONS.

'The Fijian weinmannia. The Skottsberg's wikstroemia. The Prony Bay xanthostemon. The Maui ruta tree. The root-spine palm. The Franklin tree. The Cuban erythroxylum. The fuzzyflower cyrtandra. The Szaferi birch. The Cuban holly. The Hastings County neomacounia. The Yunnan malva. The toromiro. The Mason River myrtle ... '

'Lost flora,' says the voice, and you have the sensation of a door softly closed, a latch slipping down inside. Again, you weren't aware the litany had ended. Your biro moves across the table, overshooting its mark. It occurs to you that much time has gone. You were lost in the murmuration, and when you skip back over the pages you find that your notebook is almost full. Hurriedly you fumble in your journalist's pouch in search of a replacement. Glasses clink mildly around the table. You have missed the toast again. The waiters bring the vat.

FIGURE 11: LOSS SOUP (C). The giant tureen still echoes emptily as the soup crashes into the china depths. It looks as if an ocean could slide in there. The oily smell rises unpleasantly, saturating the air around. The smell makes you uncomfortable. It's better to breathe through your mouth.

FIGURE 12: THE FOURTH INTONATIONS.

'The arctops. The sycosarus. The gorgonops. The broomisaurus. The eoarctops. The cephalicustriodus. The dinogorgon. The leontocephalus. The inostrancevia. The pravoslaveria. The viatkogorgon. The aelurognathus tigriceps.'

'Gorgonopsians,' says the voice. You don't even know what this means. You check the egg-timer timidly, shaking the cramp from your pen-clawed hand, but the sand is still flowing down, a never-ending stream.

FIGURE 13: LOSS SOUP (D). Another greyish slurry emits from the vat, frothing as it hits the china walls. You notice some of the diners' mouths are shielded with scented handkerchiefs. The stink is becoming immense.

FIGURE 14: THE FIFTH INTONATIONS.

'The Gallina. The Karankawa. The Anasazi. The Caribs. The Thraco-Cimmerians. The Lusatians. The Khazars. The Kipchaks. The Sassanids. The Olmecs. The Hittites. The Etruscans. The Babylonians. The Picts. The Fir Chera. The Gauls. The Tasmanian Aborigines. The Yeehats. The Sumerians. The Carthaginians. The Calusa. The Taino. The Ojibwa. The Mohicans. The Cahokia. The Aquitani. The Vindelici. The Belgae. The Brigantes. The Maya. The Dal gCais. The Uí Liathain. The Thracians. The Hibernians. The Kushans. The Macedons. The Amalekites. The Hereros. The Zapotecs. The Atakapas. The Zunghars. The Harappans. The Magadhas. The Moabites. The Pandyans. The Nazcans. The Timurids. The Seljuks. The Huari. The Chachapoya ... '

You are filled with a sense of despair. There appears no meaning behind these names. There is nothing to clutch onto here, they scarcely seem worth the breath they're spoken with. You halt your hopeless scribbling – already you have skipped dozens, scores, perhaps hundreds have not been committed to paper, you will never recall them now – and scan instead the line of faces seated around the dining table, pointlessly and passionlessly intoning. They have no features, no identifying markings. They have reverted to a monotype, ethnically, sexually and culturally dilute. It's as if every race in the world has been boiled down to its component paste and stirred together into a beige-coloured blandness.

In increasing journalistic desperation, you search for something, anything. Some clue as to who these people are, or more importantly, why they care. But do they care? Why are they here? You try to remember what you have heard in the past about the Dinner of Loss, but find even this has slipped away. What is this roll call supposed to be for? What are you meant to be observing?

You close your notebook, and then your eyes. You'd like to close your nose as well, but the reek of the soup is all-pervading, it's already inside your skin.

FIGURE 15A: THE EGG-TIMER (B). The silence is more general than before, and it takes you a while to understand why. The sand. The sand has finally stopped hissing. You open your eyes, and see that the diners have turned their heads to the far end of the hall, where, sure enough, the top chamber stands empty, and the bottom chamber is full.

FIGURE 15B: THE EGG-TIMER (C). More servants appear, and commence an operation that involves a set of tiny keys, which they use to loosen the brackets that hold it together. You realise the entire egg-timer unscrews, to divide the top from the bottom chamber. The empty top chamber is leant against the wall, while it takes six men to carry the bottom, staggering towards the dining table with the great sand-filled glass bell.

Somehow they lift it onto the table and then clamber onto the table themselves, dragging it to the soup tureen. Amid much grunting and strenuous groans the sand is poured into the soup, every last grain shaken out of the chamber. Then the concoction is thoroughly stirred with the oversized ladle.

The pungency of the odour mounts. The diners are gagging politely. You pull your sweater over your face and try not to breathe it in.

Finally the servants do the rounds, ladling soup into each wooden bowl.

'Ladies and gentleman, loss soup,' says the voice, with infinite sadness.

FIGURE 16: LOSS SOUP (E). You stare in some horror at what lies before you. It reeks of bilges, dishwater. An oily film slides on its surface, and when you poke it with the spoon you disturb partially suspended bands of sallow browns and greys. Occasionally a translucent lump of matter rises to the surface, slowly revolves, and then sinks back into the anonymous slop. The sand forms a silt at the bottom of the bowl, something like Turkish coffee.

You cannot remember what you expected, but surely it was something better than this. Perhaps you imagined them swimming down there – shades of the Kipchaks, the wisents, the grebes, the canopies of long-extinct trees – but you find yourself confronted instead with a sewer-stinking broth. There's not even any wine left to wash the stuff down. Is this perhaps some awful joke?

You look around. The diners are eating, ferrying the soup from their bowls to their mouths with mute determination. The liquid dribbles from their loose lips, splashing back into the bowls. Apart from the pitter-patter of soup drops, the only sound around the table is the steady champing of teeth against sand. Throat muscles clench and gulp. They are actually swallowing the stuff.

As unlikely as it seems, you find yourself incredibly hungry. You feel as if you haven't eaten for weeks. You've lost track of how long you've been in this place. Your stomach aches with emptiness, a hunger of bottomless proportions. Steeling your nerves, you take a spoonful and bring it towards your mouth. But something tells you that would only make it worse. You just can't do it. An enormous sadness grips you. Your spoon tips and the soup splashes onto the open page of your notebook, soaking through the paper and blotting the words.

You put the notebook back in its pouch and weakly rise to your feet.

'I'd like … I'd like to add my own,' you say, holding up your empty glass. Hollow eyes swivel, but no-one speaks. 'My contribution … such as it is. I lost my father. I mean, we don't speak. We don't know who each other are anymore. And long before that, I lost a toy that wouldn't have meant much to anyone, but for me it was the only thing that seemed at all important. I left it under a tree in some woods. I used to think about it getting rained on. And … and I lost many friends. One in particular. I guess he decided he didn't see the value in our friendship anymore. I lost contact with all my old girlfriends, and even the ones I stayed in touch with, I've lost them forever too. And I lost a love that needn't have been lost. I could have kept it alive but I chose not to. And … I've forgotten certain smells and ideas. What the light was like at this or that moment, things I thought I could never forget … someone's face, someone else's name …who I was before … '

The words trail off. You've lost yourself now. Something tugs dully at the back of your mind, and for a moment you almost know what it is, but then it disappears like everything else, and you sit back in your seat.

The diners stare at you gloomily. Their jaws continue working up and down. The only sound is the sound of champing sand.

Finally you bring the soup to your lips. It doesn't taste of anything at all.

The Falling Years: An Inhumanist Vision

JOHN MICHAEL GREER

Issue 1, Summer 2010

I

Robinson Jeffers' name is hardly one to conjure with these days. The odd anthology of American poetry occasionally quotes his less troubling nature poems, and a few tourist shops in Carmel and Monterey have made a minor industry out of him, the way other towns lionise dead rock musicians or football stars. Outside of these limited circles, it's not often one hears of him.

Not until 2001 did a solid collection of his major poetic works appear – try to think of another major twentieth-century poet who was nearly forty years dead when this first happened – and The Selected Poetry of Robinson Jeffers set only the quietest ripples in motion. Gone are the days when Jeffers was so controversial that his own publishers put a note in one book of his poems distancing themselves from his views. Those who play at rebelliousness in contemporary letters might take note: make a show of iconoclasm in acceptable ways and you can count on a lasting reputation; stray into actual iconoclasm, rejecting the fashions of the avant-garde along with those of the mainstream, and the world of culture will forget you just as soon as it can.

A few details will put this extraordinary figure in his proper setting. Born in 1887, he belonged to the same generation of American poets as T. S. Eliot and Ezra Pound. Like them, he saw the facile modernist faith in progress refute itself in the cultural sterility of the Gilded Age and the crowning catastrophe of the First World War, and went in search of stronger foundations for his poetry. Eliot found his Archimedean point in a willed acceptance of Christianity; Pound, less successfully, tried to cobble together a tradition of his own from a rag-heap of sources embracing everything from Provençal minstrelsy to fascist economics. Both turned to Europe for a sense of depth they could not find on American soil.

Jeffers took a more daring approach. In the years just before the First World War, when Eliot and Pound were rising stars in a poetic galaxy rotating around the twin hubs of London and Paris, Jeffers moved to a sparsely settled stretch of the California coastline near Carmel, where he built a house and, later, a stone tower with his own hands. His quest for foundations could not be satisfied at any merely human depth, and finally came to rest in nature itself.

He called his theory of poetry 'inhumanism,' and sketched it in uncompromising terms: 'It is based on a recognition of the astonishing beauty of things and their living wholeness, and on a rational acceptance of the fact that mankind is neither central nor important in the universe; our vices and blazing crimes are as insignificant as our happiness. [...] Turn outward from each other, so far as need and kindness permit, to the vast life and inexhaustible beauty beyond humanity. This is not a slight matter, but an essential condition of freedom, and of moral and vital sanity.'

Put another way, the core of inhumanism is the principled rejection of anthropocentrism, and the pursuit of what might as well be called an ecocentric standpoint: one in which nature takes centre stage, not as a receptacle for human activities, emotions, or narratives, but as itself, on its own inhuman terms. It's an appallingly difficult project, difficult enough that Jeffers himself couldn't always sustain it; critics have pointed out the places in Jeffers' verse where poetry gives way to lecture, or descends into an inverted sentimentality that wallows in images of suffering and despair. When Jeffers achieved the task he set himself, though, the results are stunning: for a moment, at least, the claims humanity loves to make on behalf of its own importance fall silent before a universe that was busy with its own affairs for billions of years before us and won't take the time to notice our absence when we are gone.

Jeffers is thus among the few figures in literature to grasp the core feature of the universe revealed by Darwin and his successors, the perspective that the late Stephen Jay Gould called 'deep time' – the sense of human existence as an eyeblink in the long history of the planet. His answer to the spread of suburban sprawl over his beloved Carmel Point is typical:

> It has all time. It knows the people are a tide
> That swells and in time will ebb, and all
> Their works dissolve. Meanwhile the image of the pristine beauty
> Lives in the very grain of the granite,
> Safe as the endless ocean that climbs our cliff. – As for us:

We must uncenter our minds from ourselves;
We must unhumanize our views a little, and become confident
As the rock and ocean that we were made from.

As poetics, this is hard enough. As a programme for any more pragmatic engagement with the world, it poses a staggering challenge. Jeffers didn't shy away from the places where poetics and politics intersect; Shelley gave him a sense of the poets' role as the world's unacknowledged legislators, and he addressed the political arena directly in such poems as 'Shine, Perishing Republic' and 'The Day is a Poem'. Still, his politics – like his poetics – found few listeners. Most of the few critics who discussed his work at all, slid past the complex political vision that frames much of Jeffers' work with a few comments about 'isolationism', and maybe a nod to Spengler and Vico. Jeffers' prophetic ear was exact, but no one else was listening:

There is no returning now.
Two bloody summers from now (I suppose) we shall have to take up
    the corrupting burden and curse of victory.
We shall have to hold half the earth; we shall be sick with self-disgust,
And hated by friend and foe, and hold half the earth – or let it go, and
    go down with it.

Still, Jeffers knew as well as anyone that the legislation of poets needs time to have its effect. The rising spiral of environmental crises shaping today's headlines marks, I have come to believe, the point where Jeffers' vision becomes an historical fact, and his inhumanism a centre of gravity toward which any meaningful response to the predicament of industrial society must move. In saying that, I'm not claiming that responses to our crisis ought to move toward inhumanism; I'm saying that they will do so, even if those who think they are defending the environment have to be dragged kicking and screaming along that route.

I say that with some confidence because most of the journey has already happened. The anthropocentrism that runs through the environmental movement, even, or rather especially, among those who most bitterly condemn humanity and all its works, seems to me to mark a final, frantic attempt to cling to the illusion of a human-centred cosmos. As today's environmental narratives join the ruins of earlier lines of defence in history's compost heap, it's not easy to imagine any place where anthropocentrism can stake a further claim against the massed inevitabilities of nature. At that point Jeffers' inhumanism offers a glimpse at the foundations on which human thought will have to rebuild itself.

II

The environmental movement as a social phenomenon still awaits its historian, though there have been capable histories of the ecological ideas that have inspired it. A first approximation, though, shows three overlapping periods of environmental activism, each with its own distinct narratives and purposes.

The first was the period of recreational environmentalism, and ran from the late nineteenth-century through to the 1960s. Environmental rhetoric in this period focused so tautly on the value of nature as a recreational resource, that its opponents, without too much inaccuracy, could accuse conservationists of simply wanting the Government to subsidise their vacation spots. Though it's easy to dismiss the period in retrospect, its great achievement – the invention of the national park concept and its deployment over much of the industrial world – marks a historical watershed of some importance. For the first time since the felling of the old Pagan groves, the Western world recognised the point of setting aside space for nature on its own terms.

The second phase, from the early 1960s through to the 1980s, was the period of sentimental environmentalism. The spark for the transition was Rachel Carson's epochal Silent Spring, which brought extinction out of scientific journals and into the public sphere. The results shared far too much with the rest of the popular culture of the time to accomplish much – the baby seals whose Holly Hobby faces made them the mascot of the movement, for example, received far more attention than many more substantive issues – but the underlying shift in awareness is worth noting. For a significant number of people, feelings of loyalty and love once fixed firmly within the human sphere widened to embrace nonhuman nature.

The third phase followed promptly. The first stirrings of apocalyptic environmentalism appeared while the age of sentimental environmentalism was barely underway, and once it worked its way out of the fringes it quickly borrowed the same durable tropes about the end of the world that proved their appeal in other contexts. The last two decades have accordingly seen all the usual changes rung on the theme of an imminent Judgement Day, with Gaia pressed into the role more usually filled by an avenging Jehovah.

Surf the web or visit a bookstore and the resulting sermons may be found without too much effort. Alongside claims that a future of ecological horror – sinners in the hands of an angry biosphere! – can be averted if we renounce our wicked ways and get right with Gaia, you can find claims that it's already too late and the wrath of an offended planet will turn sinful humanity into so much compost, upon which the righteous remnant will presumably plant the organic gardens of the New Green Jerusalem. Gospels backstopping these sermons with a giddy range of dubious historical mythologies have flooded the market at nearly the same pace.

It's crucial to recognise the hits as well as the misses of apocalyptic environmentalism. Many of the issues that underlie claims of imminent Ecogeddon are quite real, though some have been exaggerated to the point of absurdity. Where these narratives fail is in forcing the ecological crisis into anthropocentric narratives that falsify far more than they explain.

The function of apocalyptic myth, after all, is to console the unimportant by feeding them fantasies of their own cosmic significance. It's thus no accident that, for example, the seed-times of apocalyptic ideas in Judaism have been epochs when Jews were a powerless minority whose beliefs and hopes were of no concern to anyone but themselves, just as the apocalyptic strain in today's Christianity clusters in the regions and classes of the industrial world that were most heavily marginalised during the era of so-called 'globalisation'. The environmental apocalyptic narrative is partly a reaction to the impact of deep time on our collective sense of self-importance: faced with a planetary history in which geological forces and mass extinctions hold the important roles, we've tried to claim the role of a geological force and a cause of mass extinctions.

That probably couldn't have been avoided. Like the phases before it, apocalyptic environmentalism inevitably got tripped up by the anthropocentricity it tried to escape. Recreational environmentalism reached for the insight that we owe nature space of its own, and fell back to thinking of nature as a resource for outdoor holidays. Sentimental environmentalism reached for the more challenging insight that we owe nature the same bonds of love and loyalty more usually applied to family, community, and nation, and fell back to thinking of nature as a resource for emotional indulgence.

Apocalyptic environmentalism, in turn, reached for the most challenging insight of all: the recognition that we owe nature our existence, and could follow the dodo and the passenger pigeon into extinction if we mess up our relations with the rest of the world badly enough. Like its predecessors, its reach exceeded its grasp and it fell back to thinking of nature as a resource for narratives that celebrate the supposed uniqueness of humanity just as obsessively as ever. Portraying humanity as the uniquely destructive ravager of nature, after all, is just as anthropocentric as portraying it as the uniquely creative conqueror of nature. The resemblance between the concepts is not accidental; like a spoiled child who misbehaves to get the attention good behaviour won't bring, we're willing to see ourselves in any role, even the villain's, as long as we get to occupy centre stage.

III

Still, talking about the anthropocentric obsessions of today's ecological thought in general terms is less helpful than catching sight of those obsessions in their native habitat, in the collective conversation that shapes our world. Nothing is as easy as denouncing an abstract representation of a habit of thought on which one's thinking continues to be based. Think of the way that 'dualism' was all but burnt in effigy a few years back by a flurry of liberal religious writers who insisted that all religions without exception are either dualist or non-dualist, and dualism is absolutely evil while nondualism is absolutely good!

No doubt we'll shortly see a critique of anthropocentrism along the same lines: arguing, perhaps, that the habit of anthropocentric delusion is what sets our species apart from the rest of nature and marks us out for some uniquely tragic destiny or other. Thus it's important to get past the label and examine specific ways that anthropocentrism distorts the response of today's environmental movement to the incoming tide of ecological crisis.

Compare the recent and continuing furore over anthropogenic climate change to the more muted response to the rapid depletion of the world's remaining petroleum reserves, and one such distortion stands out clearly. Both these problems are unquestionably real; both were predicted decades ago, both could quite readily force modern industrial civilisation to its knees, and both are already having measurable impacts around the world.

Yet the response to the two differs in instructive ways. Anthropogenic climate change has become a cause célèbre, splashed across the mainstream media, researched by thousands of scientists funded by lavish government grants, and earnestly discussed by heads of state at summit meetings. Nothing is actually being done to stop it, to be sure, and most likely nothing will be done; not even the climate campaigners who urge drastic action in the loudest voices and most extreme terms have shown much willingness to accept the drastic changes in their own lives that would cut carbon dioxide emissions soon enough to matter. Still, the narrative of climate change has found plenty of eager listeners around the world.

None of this has happened with peak oil. The evidence backing the claim that the world has already passed the peak of petroleum production and faces a future of declining energy and economic contraction is every bit as solid as the evidence for anthropogenic climate change; the arguments opposing it are just as meretricious; its potential for economic and human costs is as great, solutions are as difficult to reach, and it can feed apocalyptic fantasies almost as extreme as those that have gathered around climate change. Still, no summit meetings are being called by heads of state to discuss the end of the age of oil; there has been no barrage of mainstream media attention concerning it, and precious few government grants. Climate change is mediagenic; peak oil is not.

A core difference between the two crises explains why. Climate change, as a cultural narrative, is a story about human power. We have become so almighty through technological progress, the climate change narrative argues, that we threaten the Earth itself. The only limits that can prevent catastrophe are those we place on ourselves, since nothing else can stop us; and even our own efforts might not be enough to stand in our way. It's nearly a parody of the old atheist gibe: to prove our own omnipotence, we've made a crisis so big that not even we can lift it out of our way.

Peak oil as a cultural narrative, on the other hand, is not a celebration of human power but a warning about human limits. At the core of the peak oil story is the recognition that the power we claimed was never really ours. We never conquered nature; we merely stole some of the Earth's carbon and burnt our way through it in three short centuries. All the feverish dreams and accomplishments of that era were simply the results of wasting a vast amount of cheap fuel. Now that the easy pickings are running out, and we have to think about getting by without half a billion years of stored and concentrated solar energy to burn, our fantasies of power are proving unexpectedly fragile, and the future ahead of us involves more humility and less grandiosity than we want to think about.

One rich irony here is that the limits imposed by peak oil are, among other things, limits on our power to destroy the world via climate change. Conventional petroleum production peaked in 2005 and has been declining since then; unconventional petroleum production, even if it recovers from the slump following the crash of 2008, will tip into decline well before 2015; natural gas is on schedule to reach its peak by 2030 and coal by 2040.

As those peaks pass, fossil fuel consumption will decline, not because we want it to decline, but because our ability to extract fuels from the ground runs into geological limits. This awkward reality has not found its way into the climate change debate; nor will it, until the anthropocentric foundations of that debate are seen for what they are.

The same point can be made even more forcefully of the greater irony that surrounds the climate change debate: the fact that the shifts in global temperature painted in doomsday terms in today's media are modest, in scale and speed, compared to some which Earth has experienced many times before. Since the beginning of the Pliocene epoch some 10 million years ago, Earth's climate has been in a phase of severe cooling, and for four-fifths or so of the time that life has existed on this planet global temperatures have been far warmer than the IPCC's worst case scenarios imagine. When the Earth's climate is normal, on this inhumanly broad scale, most of its land surface is covered by jungle, and ice caps and glaciers do not exist.

A reversion to that normal temperature would obliterate our industrial civilisation with the inevitability of a boot descending on an eggshell, and could well push our species over the edge into extinction, but the usual adjustments would soon bring the biosphere into balance, as they have after the other climate changes of the planetary past. The fact that we will not be around to see this, if it comes to that, concerns no one but ourselves.

These ironies, furthermore, have direct practical implications. While anthropogenic global warming is a real and serious problem, its consequences are subject to natural limits that current thinking, fixated on images of human triumphalism, is poorly equipped to grasp. Meanwhile, another real and serious problem – the depletion of the nonrenewable energy resources that prop up today's industrial economy and keep seven billion people alive – gets next to no attention, because it conflicts with those same triumphalist obsessions. It's no exaggeration to say that the modern world might solve the global warming crisis and then collapse anyway, because it only dealt with those of its problems that proved congenial to its self-image.

IV

Sometimes, when sleep keeps its distance in the small hours of the night, I wonder if the grand purpose for which humanity came into being was simply that Earth needed a species good at digging to pull a few billion tons of stored carbon out of the ground and nudge up its thermostat a bit. During daylight hours, I don't actually believe this; if the Earth has conscious purposes we will almost certainly never know, and if by some chance we do find out, our chances of understanding those purposes are right up there with the chance that a dust mite in Mozart's wig could have understood his music or his marital problems.

It's easy to dismiss reflections such as these as a display of misanthropy. Still, it shows no contempt for an individual to recognise that he or she isn't more important than anyone else in the world. Personal maturity begins, after all, with letting go the infantile self-regard that puts the ego and its cravings at the centre of the cosmos. It's arguably time to apply that same insight to humanity as a whole. As Jeffers wrote:

> It seems to me wasteful that almost the whole of human energy is expended inward, on itself, on loving, hating, governing, cajoling, amusing, its own members. It is like a newborn babe, conscious almost exclusively of its own processes and where its food comes from. As the child grows up, its attention must be drawn from itself to the more important world around it.

The environmental crises of the present bid fair to make that shift in attention inevitable, no matter how hard we fight to keep ourselves at the centre of our own imagined universe; and in the process most of the presuppositions of human thought will have to change. Crucially, we will be forced to come to terms with the fact that no special providence guarantees our species the fulfilment of its hopes, or even its survival. Sooner or later humanity, like every other species, will become extinct, and it's a safe bet that the history that unfolds between the present moment and that hopefully distant time will be just as sparing of Utopian dreams fulfilled as has human history so far.

This doesn't deny us the possibility of improving our lives, our societies, and our relationships with the cosmos that surrounds us; it does mean that those improvements, like everything else in the real world, will take place against a background of hard natural limits that will inevitably restrict what can be attained.

One consequence is that the faith in perpetual progress that forms the unacknowledged state religion of the modern world faces a shattering disillusionment. Progress, as we have known it, amounts to little more than the race to find ever more extravagant ways to burn cheap, abundant fossil fuels. Those fuels are no longer as cheap or abundant as they once were; in the not too distant future they will be scarce and expensive, and not all that much further down the curve of history they will be so scarce, and so expensive, that burning them to power what remains of an industrial society will no longer be a viable option.

Nor can we simply count on, as too many people are counting on, the hope that some other energy source equally cheap, convenient, and concentrated will come along just as we need it. The fossil fuels we burn so blithely today are the product of hundreds of millions of years of complex ecological and geological processes. At the dawn of our now-receding Age of Excess, they represented the single largest concentration of readily accessible chemical energy in the known solar system. Insisting that an industrial civilisation dependent on this vast surplus can thrive on the sparser and less concentrated energy flows the Earth receives from the Sun day by day — which is what most current advocates of 'sustainability' propose — flies in the face of ecological and thermodynamic reality; it's as though someone who won a huge lottery pay-off, and spent it all in a few short years, insisted he could keep up the same extravagant lifestyle with the income from a job flipping burgers for minimum wage.

Instead of fantasising about the kind of future we want humanity to have, in other words, or confusing our daydreams with our destiny, we need to start thinking hard about what kind of future humanity can afford, and taking a hard look at social habits that require levels of energy and resource inputs we won't be able to maintain for much longer. A rethinking of this kind is not optional; if we refuse it, nature will do the job for us. Ecology teaches us that every species either evolves ways to limit the burden it places on nature or suffers from limits imposed on it by outside factors, and we are no more exempt from that law than we are from the law of gravity.

At this moment in history, only a massive worldwide effort of more than wartime intensity might have even a modest chance of managing a controlled descent from industrial civilisation's extravagance to some more durable form of society. The window of opportunity for so staggering a project is narrow, if it has not already closed, and the political will that would be needed to carry it out is nowhere in sight. Thus the same sort of uncontrolled descent that ended the history of so many earlier civilisations has become the most likely future for ours. Certainly this was Jeffers' view:

> These are the falling years, They will go deep,
> Never weep, never weep.
> With clear eyes explore the pit. Watch the great fall
> With religious awe.

Still, it's precisely in the troubled years ahead of us, as our civilisation stumbles down the long broken slope toward a future that will make a mockery of our fantasies of progress and cosmic importance, that Jeffers' perspective offers its most important gifts. It's the man or woman who comes to terms with the inevitability of his or her own death that best knows how to grapple with life. In the same way, Jeffers' inhumanist perspective can be a crucial source of strength now, and even more so in time to come.

When we realise that human history is nothing unique – from nature's perspective, we're simply one more species that overshot the carrying capacity of its environment and is about to pay the routine price – we can get past the habit of wallowing in a self-blame that is first cousin to self-praise, face up to the hard choices ahead, and make them with some sense of perspective and, at least potentially, some possibility of grace. Humanity cannot and need not bear the burden of being the measure of all things, Jeffers is telling us, for a saner and stronger measure is all around us:

> Integrity is wholeness, the greatest beauty is
> Organic wholeness, the wholeness of life and things, the divine
> beauty of the universe. Love that, not man
> Apart from that.

References

The poems quoted taken from Tim Hunt, Ed. *The Selected Poetry of Robinson Jeffers* (Stanford University Press, 2002)

Melba Berry Bennett, *The Stone Mason of Tor House* (Ward Ritchie, 1966)

Arthur B. Coffin, *Robinson Jeffers: Poet of Inhumanism* (University of Wisconsin Press, 1971)

Rudolph Gilbert, Shine, Perishing Republic: Robinson Jeffers and the Tragic Sense in Modern Poetry (Haskell House, 1965)

John Michael Greer, *The Long Descent* (New Society, 2008)

Richard Heinberg, The Party's Over; Oil, War and the Fate of Industrial Societies (New Society, 2003)

Donald Worster, *Nature's Economy* (Cambridge University Press, 1994)

A Present That Can Exist

AKSHAY AHUJA *in conversation with* DMITRY ORLOV

Issue 3, Summer 2012

'The role of the man who foresees is sad one,' wrote Chamfort, around the time of the French Revolution. 'He afflicts his friends with warnings of the misfortunes they court with their imprudence. He is not believed; and when the misfortunes occur, those same friends resent him for the ills he predicted.' For about a decade, Dmitry Orlov has been a foreseer. Since the financial collapse in 2008, he has been getting an increasing amount of attention – if not gratitude – from a public that is starting to sense the fragility of the networks on which they depend. Orlov's presentation, 'Closing the Collapse Gap,' from 2006 has been downloaded several million times, and he gives presentations around the country on peak oil, social collapse, and the comparative preparedness of America and the Soviet Union for economic and political breakdown.

Orlov came to America from the Soviet Union as a teenager. After the Soviet collapse, he returned to Russia for several visits and got firsthand experience of a society where the usual supports were falling away. When he returned to America, he began to realise that the Cold War enemies had a great deal in common. After analysing the parallels, he became convinced that America would, before too long, share the fate of its superpower sibling.

His book, *Reinventing Collapse: The Soviet Experience and American Prospects* (first published in 2008, and revised and updated in 2011), discusses the shared weaknesses of the two superpowers – for example, a massive and increasingly ineffectual military – and compares America's likely resilience following a collapse to what he witnessed in Russia. On his popular blog, ClubOrlov, he writes essays – philosophical, practical, darkly comic – that explore how to prepare for the end of a lumbering and unstable system.

Several years ago – convinced of what was on the horizon, and before the real estate market collapsed – Orlov and his wife sold their Boston condominium and bought a sailboat, which is now their full-time home. When Orlov isn't sailing up and down America's East coast, he works as a software engineer and commutes within the city on a bicycle.

We meet outside Boston's enormous convention centre and walk to a pleasant Middle Eastern cafe. As we eat, Orlov talks about some of skills he is picking up, from welding to playing the mandolin (the latter is for amusement on long voyages). One of his plans involves using old shipping containers, available for almost nothing, to build a home somewhere along the coast. Twenty years from now, he says, while America is disintegrating, maybe he'll be sailing with a crew, picking up salted fish from Canada and bringing it down the coast, stopping in the Dominican Republic and getting dried pineapple, sailing it back north. Meanwhile, pop music blares out of the cafe's speakers; people at another table smile and sip coffee, and seem to be living in a different world.

AA: You started making some of these connections in the 90s. What was the spark for you?

DO: I started paying attention to economics when I was in engineering school. I wasn't very pleased with the Soviet Union. I was quite anti-Soviet. I wanted the Soviet Union to collapse. I was thinking – what's keeping it going? Because it was very dependent on grain imports and oil exports. And I didn't quite see it – but I was just watching it. I watched oil prices fall, I watched how much trouble they had coming up with the money for the imports, and how the economy was stagnating and not delivering the results that were needed to keep it going. I saw all that, and then I watched the collapse and how things fell apart. And I understood that oil was the Achilles' heel of the whole system. Somewhere around the mid-90s, I put it together. Russia at that time started recovering, and I thought, well, there are very strong similarities with the United States, so how long does the United States have?

AA: Is there a constructive way to speak about these ideas?

DO: Well, I think to talk about it is a drag. It's annoying. And it's thankless.

And I'm realising that I don't actually like to do it much anymore. There are specific things to talk about: various types of self-sufficiency, various ways that work right now that are usable and that there are pockets of people practicing. For example, I live on a boat, because it's so much cheaper. I couldn't completely own a house, outright, around here. Everybody I know has a mortgage. But the boat I own free and clear, and everything on the boat is mine, built with my own hands. That's another thing that's difficult to do with a house around here, with the permitting required.

There are quite a few people doing that. There tends to be a big separation between those thinking about the big picture and the people just living their lives the way they want, and somewhere in their mind, there is this understanding that they're doing this because the environment is just not reliable – it's too unsafe – to do the mainstream thing. They're focused on the micro-level, on their personal life and their personal adaptations. That's productive for a lot of people, but then it becomes specific to where they are and what they're doing. But there's no mass movement, and I think that that's not an unreasonable thing.

AA: Do you think a mass movement could develop when the situation changes, or do you think that small-scale answers are all that can be expected?

DO: Well, most people will deny that it's happening until it happens to them. First of all, they won't have any time to think. They'll be trying to survive day to day. And second, they'll blame themselves. Or they'll blame their luck.

AA: Or somebody else.

DO: People think there'll be a lot of scapegoating going on. But in this country – it's not a meritocracy and it never has been. People basically 'luck out' in various ways. They're born into the right family, with the right amount of money, with the right connections. Or they get the right roommate in college, the admission committee favors them. It's all worship of chance. The greatest deity in American life is not Jesus Christ, it's the goddess Fortuna and the roulette wheel. That's the temple at which Americans worship. Just about every casino in this country is bigger than the national cathedral, the biggest church in the country.

So that is the operative cult. And what that cult tells you is that it's all in the hands of chance. We draw lots. And so Americans are very placid in terms of being victimised by the system because it's all down to chance. If they think that they got a chance to play the game and they lost, then they're fine with it. From the scratch ticket buyer to the person who invests big-time in Wall Street, they're all worshipers of Fortuna.

AA: Do you see an American collapse cascading across most of the developed world? In an integrated system, is any developed place going to be much better than here?

DO: I think there will be a lot of places that will more or less stay the same, even while unspeakable horrors are unfolding in many parts of this country. I can't think of any place that's worse prepared. There are a lot of communities in the world that are far more cohesive, and safer. The problem is that you can't just go there in an emergency. You have to make a home there before you actually need it. You have to get along with them.

My favourite example: my wife's family bought a country home in Russia, and people started saying hello to them after about ten years.

And that's typical. Preparing ahead of time is definitely a good idea. Having options, having contingency plans, is definitely a good idea.

AA: You've written about finding people with resources to make the transition potentially less painful. I was wondering what those projects were.

DO: Well, what is happening, but should be happening far more rapidly, is community-level food production and preservation. Local stockpiles. Then there are communities that, right now, have pumped water. And once the water mains aren't getting supplied, they won't have access to water. So another thing is rainwater collection and purification systems. Another problem is that people will be cut off from heating fuel and cooking fuel. So passive solar and solar concentrators are very important elements. And another thing is that communities on rivers and canals and coastlines will only be supplied with the things they need from the outside if there is some sail-based transport happening.

Now, the interesting thing is that if you just start buying galvanised fittings, and wire rope, and sheet metal, and sail cloth − all of those things, they're just like buying stocks and bonds, except they will yield more. Any kind of physical inventory, from now on, especially if it's made using materials that are becoming increasingly scarce − for instance, chrome molybdenum steel. These metals have increased in price by a large amount just in the last decade, because they are depleting, along with many other strategic materials.

So anytime you stockpile chrome moly tubing, for instance − well, fifty years from now you will be making bicycles out of them, but you will have paid much less for them than that material will be worth in the future. Because even if it will still be made, it'll be very expensive. It'll probably be made in only a few places on the planet and with transportation costs going up, it may be inaccessible for that reason. Or maybe this quality of material will no longer be manufactured at all. You don't have to look for examples very far. Right now you can't buy a bicycle with the quality of steel that went into making mine, that's thirty years old.

AA: Do you find that the biggest barriers to getting investors to listen to such ideas are psychological? Presumably they've done pretty well in this system if they have surplus resources at all.

DO: Having large quantities of money in your possession causes some kind of neurological damage. I wouldn't call the problem psychological. I've known a lot, well not a lot, quite a few very rich people. And there's a problem with them. They can't imagine themselves not being rich. They just lack the imagination. And so it's sort of apocalyptic for them to think of a world where they can't call up their stockbroker, or their investment advisor, because they've shot themselves or hung themselves, or run off and hid in a bunker. And so they can't really think in very straightforward terms.

AA: To ask a depressing question: many people don't think a de-industrialised planet can support more than about half a billion people. What are some of the best and worst case scenarios in terms of getting to that number?

DO: Well, the best case scenario is it happens quickly. There's a fast die-off, and the ecosystem is saved. If you think about, all the human bodies in the planet could be packed into one cubic kilometer. So that's not very significant. That's just a dot. Now, that tiny bit of biomass, which is dwarfed by lots of other types of biomass, is destroying the ecosystem at a fantastic rate. So the faster it dwindles to a size which the ecosystem can support, the more of the ecosystem will be left. The worst case scenario is business as usual until this planet can no longer support life.

AA: A quote from your book: 'We may be hurtling towards environmental doom and, thankfully, never quite get there because of resource depletion …' This might be a mystical question, but is there some sense in which the planet will not allow itself to be destroyed? Do you have any sort of faith in that?

DO: No. We could generate gas by setting spent coal pits on fire. That will take care of the rest of the ecosystem. We could have open pit nuclear reactors using not just the spent fuel, but the nuclear weapons. We could make the whole place radioactive just for the sake of keeping the industrial systems and the military systems going a little longer. There's really no limit to human stupidity. If people set their minds to destroying this planet, I'm sure they'll manage to do it.

AA: So is collapse in some ways is a hopeful thing?

DO: Oh, absolutely. Every day that goes by that collapse doesn't happen the eventual level of population that can be sustained by the planet becomes lower.

AA: One of aspects of your work that many other writers on the same subjects don't discuss is violence.

DO: This country is already horrific in terms of violence, if you look at the statistics in terms of murder and rape and road carnage. It's much worse than most places on earth that I've visited. Compared to this country, just about every place on earth – except for the really distressed ones, like Iraq, like Afghanistan, which are like that because of what this country has done there – are tame. Very, very tame. I feel unafraid there, whereas I feel afraid in a lot of places here. People here are very rigidly controlled, and have to be monitored, spied upon, medicated into submission, in order to suppress the level of violence.

But I think what will carry people away more than violence is various kinds of disease. Here, people, if they exercise, it's at the gym. It's not by doing manual labor. Their comfort zone in terms of how warm it has to be before they get too hot or too cold is really small, because they're used to air conditioning and central heating everywhere. So once they're put into a survival situation – basically, once everyone has to go out camping whether it's winter or not, whether it's indoors or outdoors – a lot of people will basically be carried away. A few years after that, you suddenly have a population that's significantly younger and significantly healthier. And that will be a very big change. Suddenly the rules by which society operates will be quite different.

AA: Tell me about the Sail Transport Network and your plans for it.

DO: Well, there were a few things attempted in that direction. My friend Dave Reid in Seattle did it, in terms of sailing organic produce, unmotorised, to Seattle and selling it at the docks. And it was a giant hit. Everybody was really excited about it until the authorities found out. And of course you can't do things like that. There are rules about safety, about having people on the floating docks. If you've ever seen a marina, there's nothing to prevent you from falling off. Now, people do fall off and drown, once in a while, a few a year. That's considered normal. But in this country you're not allowed to have people put in an environment where they have to be competent. You have to assume that they're incompetent, and protect them from their own incompetence to do business.

So that was one problem. Another problem is that most of the facilities are owned by gigantic entities, so little sailboats are just not something that they're equipped to handle. And it was actually not very far from making money, the whole venture. Of course if you had to jump through all the regulatory hoops, then it would definitely be a money loser, along with the rest of the economy. He more or less gave up on it for now. He's still nurturing plans of building a bigger vessel, and using it for training in the meantime.

The basic realisation is: There's a before and an after in a collapse scenario. Before the collapse happens, the solutions that would work after the collapse are uncompetitive and illegal. And after a collapse happens, the solutions that would work cannot be put together because there's this thing called collapse, and nothing is moving, and nothing can be done.

So we started thinking through this process and that's where I got the idea of attracting investment capital to making plans and stockpiling parts and equipment, to then start executing on these plans *once* nothing is moving. And as a hedge against collapse not happening you still have this valuable inventory that will continue to appreciate in price, if properly stored. And I put out this idea and I've received absolutely zero response.

AA: What do you think is preventing a bigger response?

DO: Well, there are several things preventing a bigger response. First of all, it contradicts what people are being told. Secondly, the message itself creates cognitive dissonance on many levels. Thirdly, you know, who am I to spell these things out for people? I'm just an opinion in the sea of opinions. So why should anybody treat my voice as more significant and more valid than their own?

AA: How long do you think it would take for such regulations to loosen? Will it be gradual and progressive, or...

DO: It could be progressive like that, or it could be a series of riots and then the police disappear from the streets, go home, and are never heard from again. And you have a bunch of impostors in police uniforms running around. And everybody realises that they're not really interested in supporting the politicians or enforcing the existing laws. They're interested in providing for themselves and their families by using force. Then the picture changes dramatically, and all sorts of things that weren't possible before become not only possible, but very useful, because then you have something to pay these thugs off with.

AA: I think everyone's pretty skeptical about the normal political process, but you seem to have some admiration at least for the Occupy people. So I was wondering what kinds of direct political action, if any, are worth engaging in.

DO: Occupying. It's very useful. Occupy everything in sight. It's your country – take it back. If somebody thinks they owe something, convince them otherwise. And if somebody thinks they *own* something, convince them otherwise. Nobody owes anything and nobody owns anything. The whole place is bankrupt.

AA: You've written about Russia – are there any other historical examples that are instructive to look at?

DO: Well, in many ways it's unprecedented. I think parallels can be drawn between the United States, as it is now, and Nazi Germany and Japan. Because these are both situations where a country that really depended on fossil fuel imports was cut off from fossil fuel imports. The Japanese fought and lost the war trying to get to where the oil was. The Nazis fought and lost the war trying to get to the oil in Baku, in what is now Azerbaijan. Once the tide turned at Stalingrad and it was clear they were not going to get the oil, that was the end.

AA: The Internet is the main way these ideas have gotten the following that they have. I'm wondering what you think about its future.

DO: Well, there could be all sorts of clampdowns on the Internet. There was an uproar in the US against SOPA and PIPA, and those got withdrawn. For now. But at the same time the EU accepted very similar laws, without much debate at all, because the EU at this point is so opaque. And also the US government went after Megaupload and shut it down without SOPA and PIPA, just using some other existing laws (or ignoring some existing laws, which is something they do as well). As our message becomes impossible to ignore, I expect that it will be suppressed to a large extent. I think the funny thing is that that won't make any difference.

AA: Suppressed by blocking the websites?

DO: Yes. But I don't think it will make any difference for two reasons. One is: very few people are listening to us anyway. The idea that this is going to spread like wildfire and everybody will suddenly start downloading my stuff – forget it. It's not happening. Secondly, at some point there will be a tipping point at which the mood of the entire country will shift – when people will realise, this is not about *me*, this is not about my luck of the draw, this is by design, and if I keep playing the game, and if the people I know keep playing the game, then we're surely going to die.

And when that gets through to enough people, there will be a chain reaction that happens person to person, within families, within communities, neighbourhoods. And then it'll be impossible for anyone to stop whether they've lost the Internet or not, because the organisation will be in the form of physical bodies rather than bits flowing over the ether.

AA: A lot of people have partners who don't have quite the same ideas, and I was wondering how you and your wife arrived at consensus on what to do.

DO: I think it's normal for everyone to have a slightly different idea. My wife thinks for herself. I think we agree about a lot of fairly basic fundamental things about how life should be lived. We both detest this consumerism that forces us to buy short-lived garbage for quite a bit of money. We both don't like a society where people are running around pretending to be incredibly busy just so they feel important, whereas they're actually doing nothing productive of any sort. There's this incredible emphasis on material things, which are actually not material at all, they're ephemeral. People are fixated on bits in a computer, and trashy little consumer items. These are not material possessions of actual value. Houses with paper thin walls that are for some bizarre reason worth half a million dollars. Just in terms of how we approach life, I think we are very much in agreement.

In terms of how to get by or get along, you know, my wife has survived the collapse of the Soviet Union. I just went there to visit. First of all, she isn't looking forward to it, because why should someone go through something like that twice. Second, she knows just how unpredictable it is, so she'll only believe her own eyes, not something I say. The other thing is that, it's perfectly normal for people to want to live a good life right here and now, no matter what the future holds. It's certainly stupid to work like crazy to towards a future that doesn't exist. That's definitely insane. But working toward a present that can exist is not such a bad idea at all.

On this Site of Loss

HANNAH LEWIS

Issue 3, 2012

I

> To emigrate is always to dismantle the centre of the world, and
> so to move into a lost, disoriented one of fragments.

Even by the standards of our age, in which experiences of lostness and fragmentation have become near-universal, Rashad's life has been unusually
marked by displacement. The son of an Iraqi diplomat and his German wife,
his childhood was a sequence of sudden and profoundly disorienting shifts of
context. He remembers Nubian lullabies in Sudan and revolutionary songs at
a Catholic nursery in Beijing, though he has forgotten whatever he once knew
of these first languages. There are two memories of being shot at in a car, once
in Iraq, in the Baathist revolution of 1963, and later in the Libyan revolution of
1969. In between, a quieter period in Sweden: skating on a frozen lake; a barn
stacked with tools dating back centuries, and walk-in closets full of Victori-
ana; once, he alarmed the family's guests by brandishing a musket at them.
He returned to his father's homeland in his teens as a stranger, ill at ease in
yet another unfamiliar culture and initially refusing to learn Arabic; yet as he
entered adulthood, Baghdad became one of his first loves. The devastation she
has suffered since is the most traumatic of all the dismantlings of his world.

Among these formative journeys, the most extraordinary took place not on
land, but at sea. Between November 1977 and April 1978, as a 20-year-old
art student, he was a crew member and Arabic interpreter for Thor Heyerdahl's Tigris expedition – a role that included explaining the process of
boat-building to the team of Marsh Arabs who constructed the vessel, in
collaboration with Aymara Indians from Bolivia, whose reed bundle boats
were closest in design to the ancient Sumerian craft Heyerdahl sought to
emulate. The Tigris raft travelled from the marshes of Iraq, down the Gulf

and across the Arabian Sea to Pakistan, then back across the Indian Ocean towards Africa. Covering 4,200 miles, this small yet resilient craft offered persuasive evidence of connection (by trade and perhaps also by common cultural origin) between three of the oldest known centres of civilisation, in ancient Sumer, the Indus valley, and the Nile valley.

Rashad's reflections on the expedition today are mixed. While Heyerdahl's ideas are academically controversial, Rashad loyally defends his work, at least in the case of the Tigris. Its success, in proving that ocean travel (steering with rudders and directional sailing, tacking against the prevailing winds) was possible with the materials and technologies available to these civilisations over five thousand years ago, is undeniable. Yet he is critical of what he calls Heyerdahl's 'mono-dimensional' approach, basing the vessel's design on a single technology (reed bundles), which he says 'primitivises' ancient cultures, overlooking the sophistication and range of technologies that existed at the time.

He also has mixed feelings about the journey's status as 'part-showbusiness'. Filmed by the BBC and shown internationally, the Tigris (like the earlier Kon-Tiki and Ra expeditions) embodied Heyerdahl's flair for making ancient history accessible and gripping for huge audiences; this is part of the value of his work, yet it also shaped what was possible aboard the Tigris in ways Rashad didn't always like. The Tigris's dramatic end – set ablaze by its crew off the coast of the tiny republic of Djibouti – was, Rashad says, more a matter of 'self-interest' by a portion of the crew than the impassioned act of protest Heyerdahl described. (Wars and civil strife in Yemen, Somalia and Ethiopia had made Djibouti the only place where the boat could safely land; Heyerdahl's decision to burn it there was accompanied by an open letter to the UN Secretary General protesting the 'insane reality of our time' in which Western and Soviet arms dealing had equipped these conflicts with their lethal power.) Rashad recalls how he and other younger crew members were unhappy with the decision to destroy the vessel that had been their home for five months; they urged Heyerdahl to let them press on further up the Red Sea, or even to navigate southwards around Africa; yet the expedition's older and more successful participants had profitable careers as industrialists or TV stars to return to, five months away had already been too long. To leave the vessel in the hands of the junior members of the hierarchy would undermine their status and authority over the journey's story, so destroying the raft became the preferred option.

Despite these ambivalent memories, the Tigris was perhaps the pivotal event of Rashad's life, and its influence on his imagination endures in many forms: the image and concept of the raft occurs repeatedly in his sculptural work; the evolution of technologies and cultures is an ongoing preoccupation of his research; and the conviviality of its multinational crew remains for him a benchmark of the principle espoused in Heyerdahl's letter to the UN, 'that no space is too restricted for peaceful coexistence of men who work for common survival.'

By the time I met him, Rashad's studio in Streatham Hill had become a kind of lifeboat, an inhabited artwork, a multi-layered personal raft imbued with the longing for a home that could accompany the exile. The destruction wrought by the invasion and occupation of his homeland of Iraq weighed on him, sometimes to the point of hopelessness; yet there were also bouts of visionary intuition, in which he began to populate his landlocked raft with projects that sought to preserve what could be salvaged of the cultural heritage of collapsing civilisations. In a time of catastrophe, as for Noah – or Utnapishtim in the epic of Gilgamesh – home must be an ark.

II

Home was the centre of the world because it was the place where a vertical line crossed with a horizontal one. The vertical line was a path leading upwards to the sky and downwards to the underworld. The horizontal line represented the traffic of the world, all the possible roads leading across the earth to other places. Thus, at home, one was nearest to the gods in the sky and to the dead in the underworld. This nearness promised access to both. And at the same time, one was at the starting point and, hopefully, the returning point of all terrestrial journeys.

October 2010. I've been flicking through John Berger's book And our faces, my heart, brief as photos; these words about home are echoing in my mind. Now I'm on the phone to Rashad in Ghana; he's telling me about the new series of sculptures he is making there. They are rafts, built from the discarded objects he has found around the woodyard where he's working: scrap-wood, mortars for grinding grain, pots and pans. Each raft has its own name, a role, a personality: Rudder, Navigator, Alchemist, Office, Mama, Elephant Child, Old Man. What they have in common is their form: a platform and a mast, two lines, the horizontal and the vertical.

'What can grow on this site of loss?' Berger asks, referring to the home-lessness of the immigrant. Such loss, he suggests, is only the extreme form of an alienation which has been the general and widespread experience of modern societies.

He offers two answers, 'two new expectations – offering the hope of a new shelter'. First, romantic love: 'in the modern sense,' he writes, this 'is a love uniting or hoping to unite two displaced persons.' The second, he calls an historical expectation: not that we can return home, exactly, or imagine a return to a world in which every village was the centre of the world. Rather, 'the one hope of recreating a centre now is to make it the entire earth.'

I find myself thinking of Berger's answers, now, as I try to tell the story of Rashad's work and the role that it has played in my life.

III

It was not a raft that brought us together, but a piano: an instrument that caught my attention from the first, because I had been told to look out for it. Let me explain.

At the time, I was preparing to open a temporary shop in the indoor mar-ket at Brixton Village, close to where I live. The Remade Work-Shop would be a place where waste materials and discarded objects could be repaired and recreated, transformed into things to use or sell.

I was also working on myself, with the help of a life coach, trying to shift some beliefs and behaviours that had been limiting me. One area I struggled with was to convince myself, at a deep level, of the need to care for my body. The way I described the habits of exercise, good diet and so on to myself felt dry and unpersuasive. The coach suggested that, instead of using words, I think of an image to represent my body and how I would like to look after it. What came to mind was a musical instrument, with the maintenance and practice required to keep it sounding beautiful and capable of a full range of expression. She sug-gested that I look out for an image of an instrument that I could keep somewhere I would see it daily, to remind me of this commitment I wanted to make.

The next day, I received a phone call from a woman I had never met, who had read about the Work-Shop and knew that I was looking for examples of creative reuse. She had been collaborating with a fellow artist in the studios where she worked, who was remaking old pianos: taking instruments that were broken, that could no longer be played, and inventing new ways to play them, reconstructing them with the help of other found and recycled objects. She gave me his name – Rashad Salim – and his number. And so, the Geo-Piano arrived in my shop, becoming the focus that everyone was drawn towards as they walked in the door, and its creator arrived in my life.

Amplifying the piano's traditional role as a social gathering-place for family and friends, the Geo-Piano was a magnetic object, a point of convergence for sounds and stories from around the world. Across its surface, fragments of an atlas were collaged with poetic rather than geographical accuracy: mountains ranged at the top, rivers and bodies of water descending to deserts and a sand-pit below. To access its open workings, other artists had contributed stray objects that would strike the strings with an array of unexpected sounds: fragile hammers made with glass beads; padded amulets mimicking those given to Iraqi children at birth for good luck; reclaimed vinyl records that produced an uneven whine as you held them against the strings, rotating on a drill-like mechanism. There were ribbon drones made from electrical tape, USB-powered fan motors on flexible stalks, even a violin bow that could sound the strings at one point where a bridge had been constructed. These objects were all carefully stored in felt-lined compartments at the place where the keyboard would once have been.

The strings themselves were tuned to five different musical scales, spanning continents and cultures. I'd list them proudly to visitors, as they joined in the improvisation: 'This part's like a normal piano, Well-Tempered; this part's Indian, Kali Raga; then a Persian scale, Maqqam Nairuz, and an Arabic one, Maqqam Rast; then Pentatonic – that's like the black notes, and it's used a lot in Japanese and Chinese music.' Rashad enjoyed telling people that this 'ethno-tuning', as he called it, had been carried out by a master piano-tuner by the name of Reuben Katz, a Jew, who kneeled side by side with him, a yeshmak-clad Arab, on the pew-like cushion that had been made especially for the piano, as they tuned the instrument together. (Without the keyboard, playing was best done from floor level.)

The story of the Geo-Piano, and the Re-Piano series to which it belongs, began with a visit to Baghdad that Rashad made in 2003, a few months after the American and British invasion. Destroyed pianos in bombed-out buildings were his memory of that trip. Later he made contact with a student at the Institute of Fine Arts, where he had studied in the 70s, who had identified at least ten dysfunctional pianos, ready to be resurrected as artworks, on the premises of the Music and Ballet School, a school founded by Rashad's father during his time as Iraq's Director-General of Arts. On his visit to Ghana in 2010, he came across old pianos everywhere, relics of the days of missionaries. The mass-produced piano was an artefact of the industrial revolution, of the height of empire; the instrument to which all others in the orchestra had to tune. So there was a subversive joy in salvaging these ruined machines and retuning them to sounds their makers never imagined.

Whether this piano met the brief of persuading me to care for my body is another question. Instead, it seemed to challenge my initial assumption that the body could be considered in isolation from the soul. I found significance in the fact that the instrument that showed up, and proved so rich with meaning for me, was not a factory-fresh model that could be kept in tune with a straightforward maintenance routine. It had a history; it had, in fact, been devastated; its mechanism had failed, the keyboard no longer functioned, and it had been dumped outside, its insides exposed, to rust in the rain. But that was not the end of the story; this broken machine and its decaying materials had become the ground for something new, improvised out of elements of the old, a hopeful synthesis.

IV

By turning in circles the displaced preserve their identity and improvise a shelter. Built of what? Of habits, I think, of the raw material of repetition ... words, jokes, gestures, actions, even the way one wears a hat.

In Turkey under Atatürk, traditional headgear such as the fez and the turban were banned for not being modern. People died protesting against this ban.

The meaning and power invested in headgear is another of Rashad's fascinations, one he can trace back to a favourite cap that he wore constantly as a small boy in China; and the Headgear Project is as much an expression of his work as the pianos.

Years of research have led him to a detailed typology, the first twelve categories of which, he suggests, reflect a chronological sequence unfolding through human history. It's impossible to be sure, since for many of the organic technologies involved, no archaeological evidence remains.

He gives me an old document about the project that includes this summary of his categories:

1. Natural head – hair and skin; grooming and manipulation, primary technical processes, e.g. matting, combing, plucking, cutting, braiding, cordage, tattooing, scarification, etc.

2. Shade/Shelter – as the bringing of the hand to the head creates, and the need for protection from the elements may provide; overhang, visor, hood, bonnet, basket, etc. E.g. Inuit snow goggles mimicking the shading of eyes by squinting.
3. Substance – powder/liquid/oil/fat/wax/gel applied to the head and/or face with associated technologies such as milling, pressing, filtering, etc.
4. Object – placed into or attached to hair or head, e.g. twig, feather, bone, stone, ties, comb, jewellery, pins, plugs and studs, beads (drilling) etc, that modify the exterior head (for spectacles and headphones see 'inner head').
5. Band – cord, braid, strap, diadem, garland, wreath to crown, a bandana though of cloth is a band, key techniques such as knot and bundle.
6. Net – knotted, woven or linked; hair, fibre, basket, sock such as: snood, balaclava, chainmail, nylon hose, etc.
7. Cap – bark, felt, cloth, metal, etc, following closely the shape of the head from held in place by the crown/upper head to attached (symbolic); e.g. yarmulke.
8. Hat – closed form fixed in a shape departing from the head; Fez is a brimless hat.
9. Helmet – protective form, soft and hard; Indigenous, Civil, Sport and Military.
10. Cloth – headwear from length of cloth (square, rectangular or length; triangular tend to band) changeable in form and function; a stitched/ fixed turban is a hat.
11. Inner Head – headgear influencing or taking over workings of the head, e.g. prosthetics like hearing aids, false teeth. Transition between the inner and outer head like implants and objects wired into or extending from the head, e.g. antenna, microphone, headphone; also, expression and manifestation of spiritual and psychological states and culture.
12. Virtual Head – the head encased in a synthetic, informational (virtual) reality, the artificial head (simulacrum and golem) and the captive or transplanted head/brain.
13. Mask – such as an object held to the face or a band with eyeholes or camouflage net mask, etc … (I shall make the case that mask can be arrived at by any of the genera and thus is a generic objective.)

A list like this offers an approach to history that yields different insights from the 'monumental' narratives we typically follow in school. Each type of head-gear is related to the development of certain technologies, and linked also with natural phenomena that people observed and from which they learned; thus the evolution of headgear represents the unfolding of human consciousness, our awareness of what is around us, and a process of co-evolution with our environment. By contrast, Rashad suggests, the stories we grew up with were skewed by their dependence on archaeological remains, while overlooking those technologies and behaviours which leave little trace, because the materials were organic or the techniques ephemeral.

The archaeologist Walter Andrae, who took part in the excavation of Babylon and other sites in Iraq in the early twentieth century, argued for the existence of earlier societies that were highly developed, yet left behind no stone monuments or metal tools, relying instead on forms of architecture and technology in which reeds, wood and other organic materials featured prominently. But in his time, it was to the advantage of the European colonial powers to define human history as a sequence of centralised empires and their monumental architecture; so the archaeology of monuments was promoted, while narratives of 'organic civilisations' and their technologies were suppressed or ignored.

The investigation of such technologies cannot help but be speculative, drawing on a kind of empathic attunement to our ancestors' world of experience; its imaginative leaps may be grounded and expanded by practical reenactments like Heyerdahl's voyages, which have both inspired Rashad's work and provoked him to think differently. While on board the Tigris in 1977, Rashad became convinced that the original vessels would not have carried so much dead weight as Heyerdahl's reconstruction (the 18 by 6 metre vessel weighed 80 tonnes). The surviving images of reed-boats show only the outside of the vessel, and the Tigris was built to mimic this appearance, yet what would originally have been inside remained a mystery. Certainly such boats could not have carried buffalo across oceans – as depicted on cylinder seals of the time – if they had been constructed like the Tigris, with a solid mass of reed bundles. Rashad argues that vessels built for long-distance voyages of discovery would – like the spaceships of our own time – have drawn on every kind of technology available, and catalysed the development of new ones.

Today his vision is to coordinate a multi-stranded 'harvest' of produce and techniques from across the Mesopotamian watershed (Iraq and neighbouring areas), working with universities, civil institutions and communities of craftspeople to construct and send boats of all sizes along ancestral trade routes down the many tributaries that lead into the region's great rivers, the Tigris and Euphrates. During their making and along their journeys, these boats will gather knowledge on every indigenous and low-energy technology available in the region – from weaving to pole lathes to cheese-making – using both digital means and ancient methods such as cylinder seals to document and communicate each technique, disseminating the means to build self-sufficient, off-grid local economies in communities devastated by conflict and resource shortages. At the confluence of the rivers, he envisages applying the combined knowledge of the converging vessels to the shared challenge of building an 'Ark for Iraq': a re-imagining of what might have been built to survive a prehistoric flood. Rashad's vessel would stand in contrast to the colossal wooden ship of western Noah myths; 'faced with signs of an approaching cataclysmic deluge and flood,' he argues, 'it is improbable that any people would construct an unknown, unproven vessel for sanctuary ... a prehistoric Mesopotamian community would more likely have assembled a gathering of available watercraft using local materials and craft techniques.' The boats that make up this proposed Ark are known from ancient iconography and texts, and have remained in use in Iraq until recent decades: among them the 'guffa' basket-coracle – revived from near-extinction in 2016 through Rashad's work with craftspeople in Babylon – the reed-bundle 'shasha' and 'zaima', the 'kelek' raft and 'tarrada' canoe. Brought together into a pattern of order and unity, they form an Ark that symbolises and contains the collective knowledge of a community rooted in its environment, rather than the exceptional vision of an individual.

Such a project lies in a liminal zone between science, history and art; where evidenced knowledge of the past is added to and altered by reasoning and by acts of imaginative manifestation in the present, experienced as much intuitively as intellectually. The Tigris and its successor expeditions, like the Ark for Iraq and the Headgear project, point towards a re-inhabiting of ancestral knowledge; a rediscovery – however tentative – of something of what was lost along the way.

V

When we met in 2009, Rashad's everyday headgear was a baseball cap with a yeshmak (keffiyeh) tied around it. He called it a piece of performance art, which he had been performing since September 2001. In fact, he had been wearing the same combination since the previous year, but it was only when Islamophobia became an issue and people started reacting strangely to the *yeshmak* that he began to consider it an artwork. It's the adaptability of the scarf – a simple square of fabric that can be worn in countless ways on the head and other parts of the body – that he loves, as an embodiment of democratic design, expressive of the individuality of each wearer. He sees an odd irony in American propaganda against 'towel-heads', which treats it as a symbol of ideological conformity.

This headgear was not his only response to the events of 11 September, 2001. There was also the Inverse Globe: an unmade monument, or a novel we might write together, about what could have happened if the response to those attacks had been other than it was, a turn towards global peacemaking rather than an escalation of warmongering. At the heart of the story is an image; the closest I've found to a physical embodiment of Berger's proposition that 'the one hope of recreating a centre now is to make it the entire earth.' Rashad envisioned this image as a monument to be constructed at Ground Zero.

The monument was an Inverse Globe: the Earth turned inside out. As the novel begins, ten years after the 2001 attacks, a group of global leaders and influencers stand on the platform at its centre. All around them, projected onto the interior of a great spherical building, is the image of the Earth's surface from space. Through the workings of perspective, it appears almost identical to a convex globe; but we, the observers, are within it rather than outside. We have only to turn our heads to take in the entire world at a glance. Satellite imagery and statistical information allow all kinds of data to be overlaid onto this global map: temperatures, weather patterns, deforestation, ice melt, extinctions, drought, war, resource production and depletion. This data-centre functions both as public monument and tourist attraction, but also as the hub for a global network of scientists and innovators, seeking to restore the Earth's damaged ecology to health and resilience.

When Rashad told me about the Inverse Globe, it reminded me of Stewart Brand's vision of the Earth from space, the vision which inspired the Whole Earth Catalog. Brand was certain that this glimpse of the wholeness, uniqueness and vulnerability of our home planet would cause a radical breakthrough in human consciousness. Rashad seemed equally sure of the efficacy of the Inverse Globe. And, I thought, equally over-optimistic.

Still, I have rarely seen represented with such clarity, even if only in imagination, the desire to make the entire Earth the centre of one's world – and thus, home.

VI

Why is it that so much of his life's work should feel like my own?

There is, in the first place, the romantic love which Berger describes: 'a love uniting or hoping to unite two displaced persons.' Yet how can I, who haven't moved more than thirty miles from my parents' home since leaving it thirteen years ago, think of myself as a displaced person – especially in relation to a man who had lived in eight countries on four continents by the time he was my age? Why do I identify so strongly with his sense of displacement? One trauma of my early life might bear a weight equivalent to those multiple dislocations. When I was four years old, my father committed suicide. It left me with a permanent sense of being not-at-home in my surroundings. As a child, I always felt an outsider; as an adult, I have become aware that I am far from alone in this. As Berger writes, 'the homelessness, the abandonment lived by a migrant is the extreme form of a more general and widespread experience. The term "alienation" confesses all.'

Sharing in this general alienation, Rashad and I are nothing unusual. But for me the loss of a father, and for him the early lack of a consistent home – now compounded by the devastation of his father's homeland – seem to have conjured a persistent and intense longing for the wholeness and home-ness we have missed. For him, this manifests in these projects of research and art that bring together a broad sweep of cultural and technological artefacts from various stages of human history into new and mongrel forms of beauty, as if to say: all this past remains present; all these places are here.

My own dislocation reached crisis point at the age of twenty. The move from home in Slough to college in London had been slight, in physical distance, but the inner disorientation was absolute. Berger's account of the migrant experience describes this precisely: 'Not only ... leaving behind [home, and] living amongst strangers, but, also, undoing the very meaning of the world and – at its most extreme – abandoning oneself to the unreal which is the absurd.' The framework of interpretation on which I had relied in childhood was thoroughly dismantled by exposure to postmodernist thought, or the half-digested fragments of this one swallowed as a design student at Goldsmiths around the turn of the millennium. Yet no stable foundation of meaning emerged in its place; and meanwhile, my sense of personal identity was challenged by the sometimes brutal social world of a similarly lost flock of young adults, as well as by my own determined quest to demolish and reinvent parts of myself I didn't like.

Unexpectedly, as this sense of displacement reached its peak during my second year at university, came some of the most striking moments of feeling at home in the world that I have known: a spell of altered states of consciousness, spontaneous and not drug-induced. These included mystical experiences when the physical world felt illusory and it seemed that all of time and space were gathered into a single point in which everything was present: the ultimate home, the place of convergence; a state of consciousness I later recognised evoked in the image of the Inverse Globe. I wasn't sure how I had got there. Later I began to see this as perhaps the only possible resolution of my previous disorientated state. Incongruity and contradiction between the various narratives by which I'd explained and justified things reached an extreme where they suddenly annihilated each other: the tottering edifice of stories collapsed, leaving a kind of inner Ground Zero. I was surprised to find there a feeling of bliss and faith; I couldn't hold onto it, yet still feel its resonance.

Another sudden insight shook me with the peculiar impact of a thought that reclassifies all other thoughts – the realisation that ideas evolve. Until then, I had thought that ideas might be true or not true, or more or less true, in relation to a more or less static world. Suddenly it was clear to me that ideas and stories developed in relation to each other, to the whole surrounding ecology of ideas, practices, and interpretations of experience, which might propagate, mutate, conflict with or override one another.

During the decade since, beneath the other preoccupations that have kept me busy, this notion of the 'evolution of ideas' has remained in my awareness like a seed waiting to germinate; awaiting a context in which to develop and be articulated. In the subtle evocations of the unfolding of human culture represented by the Geo-Piano and the Headgear Project, I recognised a potential opening for this articulation.

Most of all, though, it was my own experience of dislocation that I was shocked to recognise in the Geo-Piano, or rather in the story of the piano as it had been before its re-making, dumped outside with its strings exposed. The telling feature was the missing keyboard, the broken command-and-control mechanism; this was, to me, a startlingly exact metaphor for the way the fixed mental connections that had once predictably governed my ideas and behaviour had become severed – in a way that threatened to reduce me, like the piano, to permanent silence, since without them there was no established way to communicate. The reinvention of the piano as an open theatre of ad hoc assembly and musical improvisation, drawing on the sounds of other times and places, was for me an evocation of the possibilities for creative regeneration in my own life; a pointer towards how I might find my voice again.

What if, it seemed to propose, the way to a new music was not to recon-struct the familiar mechanism, but to dare to engage directly – hands-on or equipped with whatever came to hand – with the strings, the source of the sound? Not to attempt to re-tune to the original scale, but to welcome in a different set of musical intervals; to allow the interplay of these to unfold, as it naturally would, in complex polyrhythms, through phases of discord and concord, and harmonies that might not be classifiable as one or the other.

The quotations which open sections 1, 2 and 4 are taken from John Berger, *And our faces, my heart, brief as photos*.

TWO

We reject the faith which holds that the converging crises of our times can be reduced to a set of 'problems' in need of technological or political 'solutions'.

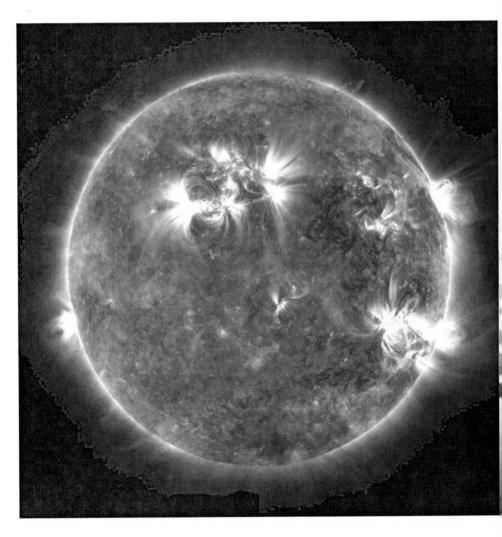

MARNE LUCAS AND JACOB PANDER — Sun from Incident Energy — *Black and white infrared 4-channel video stills (trt: 20 minutes)* — *Issue 8. Autumn 2015*

Incident Energy is a multi-channel video filmed with thermal imaging (IR) cameras exploring themes of nature, culture and the body. Using modern dancers to express a creation story the film examines energy, the universe, love, birth, conflict, decay and death alongside our (c)overt acceptance of electronic surveillance. Highly sensitive infrared surveillance technology frames the luminous energy of the body to reveal hidden heat signatures. Intelligent movement, human emotion and natural landscapes are revealed by infrared surveillance technology, juxtaposed with vast galactic imagery (Sun image courtesy of NASA/SDO and the AIA, EVE and HMI science teams).

The Shuttle Exchanged for the Sword

WARREN DRAPER

Issue 2, 2011

> Chant no more your old rhymes about bold Robin Hood
> His feats I but little admire
> I will sing the Achievements of General Ludd
> Now the hero of Nottinghamshire
>
> — Anon, from the files of the Home Office

If you should someday visit York Castle, you will encounter the romanticised celebration of the life of one John Palmer, a violent thief and murderer whose early exploits include the pistol-whipping and torture of a 70-year-old man, and the aiding and abetting of the rape of two women. Palmer is better known today for his later exploits, and by his real name: Richard Turpin.

The Dick Turpin of popular imagination was formed by Richard Bayes' semi-fictional biography, The Genuine History of the Life of Richard Turpin, the noted Highwayman. This account of Turpin's exploits was hurriedly published in 1739, shortly after Turpin was hanged for horse-theft. Biographies of condemned villains were popular in the 17th and 18th centuries and a known murderer and horse-thief was considered the worst criminal imaginable.

Bayes' described Turpin as a highwayman, but this was not his reputation in his own lifetime (when he was better known as 'Turpin the Butcher'). This encouraged later authors to further embellish Turpin's exploits. In his 1834 novel, Rockwood, William Harrison Ainsworth gives Turpin his legendary horse, Black Bess; in fact, a borrowing from a famous story of the highwayman John Nevison, who was said to have ridden the 200 miles from Kent to York non-stop, in order to establish an alibi for a crime he had committed. In Ainsworth's novel, Turpin has Black Bess gallop overnight from London to York, where his beloved mare promptly dies of exhaustion. The story of Turpin and Bess was so well received that it became popular fodder for the Penny Dreadfuls of the 19th century, and thus a powerful (but inaccurate and exaggerated) legend was born.

On 16th January 1813, some three quarters of a century after Turpin was hanged, another 14 lives would end on the gallows at York Castle. The crimes for which these young men died would be recorded variously as 'riot, breaking and entering and attempting to demolish William Cartwright's water mill (for finishing cloth by machinery)'. Unlike Turpin, their actions were motivated by more than greed – indeed, it was the greed of other men that sent them to their deaths. But no gravestone, plaque or waxwork exhibit marks the passing of their lives. Instead, their legacy is a shallow, overused and inappropriate insult thrown around by champions of the myth of inevitable, beneficent industrial progress.

Today, the word Luddite has become one of those tame insults which are used to suggest the superiority of the user without being deemed overly offensive to the intended recipient. In his 'Tips for Transhumanist Activists' (transhumanists study and promote opportunities for enhancing the human organism, and thereby the human condition, through the use of technology), Michael Anissimov says:

> Don't use harsh, insulting, unkind words to describe people who disagree with your views ... Using words like 'stupid', 'ignorant', and 'daft' smack of elitism, and reflect negatively on the speaker, only making it clear to everyone that their brain is firmly stuck within the pathology of name-calling and tribalistic thinking. If we must use some sort of adjective to describe the people we think are our 'opponents', then 'Luddite' should do.

The implication is that the Luddites were not only opposed to new technology, but were 'stupid', 'ignorant' and 'daft' to hold such a position. This attitude is not unusual: even academics are prone to treating the historical Luddites as little more than a naïve and reactionary backlash against the inevitability of change. As Eric Hobsbawm observed 60 years ago:

> [A]n excellent work ... can still describe Luddism simply as a 'pointless, frenzied, industrial Jacquerie', and an eminent authority, who has contributed more than most to our knowledge of it, passes over the endemic rioting of the eighteenth century with the suggestion that it was the overflow of excitement and

high spirits. ... In much of the discussion of machine-break-
ing one can still detect the assumption of nineteenth-century
middle-class economic apologists, that the workers must be
taught not to run their heads against economic truth, however
unpalatable; of Fabians and Liberals, that strong-arm methods
in labour action are less effective than peaceful negotiation; of
both, that the early labour movement did not know what it was
doing, but merely reacted, blindly and gropingly, to the pressure
of misery, as animals in the laboratory react to electric currents.
The conscious views of most students may be summed up as
follows: the triumph of mechanization was inevitable.

The truth about the Luddites was very different. Far from being a naïve,
disorganised mob, they were highly-skilled independent craftsmen whose
way of life had been under attack for decades.

The story of the men hanged at York had begun a little over a year earlier,
in the village of Bulwell, four miles north of the city of Nottingham. In
the north of England, the night of 4th November is traditionally known as
'Mischief Night' because children were allowed to play tricks on the rest of
the community – much like the American tradition of Trick or Treat – and it
was on that night in 1811 that a band of men with blackened faces marched
through the streets of Bulwell to the workshop of a master weaver named
Hollingsworth, whom they said 'had rendered himself obnoxious to the
workmen.' Armed with a variety of hammers, axes, pitchforks and pistols,
the men forced entry into Hollingsworth's premises and smashed up a half-
dozen wide-lace-frames – new machines which were said to do the work of
many men in a fraction of the time.

Not content with their actions, the men returned on the following Sunday
night to finish the job, but this time Hollingsworth had a team of gunmen
lying in wait. A young weaver by the name of John Wesley (or Westley) from
the nearby village of Arnold was shot as he tried to gain entry to the premises. With his dying breath, he exclaimed, 'Proceed, my brave fellows, I die
with a willing heart!' This so enraged the mob that they pushed forward,
regardless of the gunfire. Hollingsworth's gunmen fled and the workshop
was burnt to the ground.

On the same night, other frames were destroyed in nearby Kimberley,
with similar attacks taking place throughout the surrounding areas on
the following Monday, Tuesday and Wednesday nights. On the Thursday,
the body of John Wesley was carried through the streets of Arnold by a

procession of a thousand men. They were met by six armed magistrates, a company of mounted Dragoon guards, a local militia, and a posse of volunteer constables. One of the magistrates read the Riot Act, whereby people were supposed to disperse under threat of immediate arrest. The procession ignored him, and after a scuffle many of them were taken into custody. But if the authorities thought that this might put an end to the matter, they couldn't have been more wrong; Nottinghamshire erupted.

Despite the presence of thousands of troops, over 800 looms were broken in Nottinghamshire during the final months of 1811. And each act of resistance, so it was claimed, had been conceived and conducted under the order of one man. As Kirkpatrick Sale tells the story, in Rebels Against the Future:

> It was now that anonymous letters explaining the causes of the machine breaking and threatening more of it started appearing throughout the district, mailed to or slipped under the doors of hated hosiers, sent to local newspapers, or posted in the night on public boards – 'many hundreds' of them ... reported one manufacturer. All announced that a new concerted movement was afoot; all were signed by, or invoked the name of Edward (Ned) Ludd, 'King,' 'Captain in Chief,' or 'General.' ... Luddism had begun.

There has been much speculation as to the origins of the name, but the most popular theory also seems to be the most likely. Edward Ludd (or Ludlum) is said to have been a boy from the village of Anstey, just outside Leicester. In 1779 he was employed as a weaver's apprentice, but was beaten for 'idleness' by his master. Not much caring for being whipped like a dog, young Ned took a hammer to two of his master's knitting frames and 'beat them into a heap'. From then on, whenever anyone damaged a loom, whether by accident or with malicious intent, it was common to say, 'Ned Ludd did it'. We cannot know the truth behind this story, but we do know that 1779 marked the beginning of the end of the traditional weaving trade. In the closing years of the 18th century, the weaver's way of life came under threat, not only from the introduction of new technology, but also from the newly emerging capitalist approach to production. Sale depicts this vividly:

> Lancashire, say 1780:

> The workshop of the weaver was a rural cottage, from which when he was tired of sedentary labour he could sally forth into

his little garden, and with the spade or the hoe tend its culinary productions. The cotton wool which was to form his weft was picked clean for him by the fingers of his younger children, and was carded and spun by the older girls assisted by his wife, and the yarn was woven by himself assisted by his sons. When he could not procure within his family a supply of yarn adequate to the demands of his loom, he had recourse to the spinsters of his neighbourhood. One good weaver could keep three active women at work upon the wheel, spinning weft [although] he was often obliged to treat the females with presents in order to quicken their diligence at the wheel.

Lancashire, say 1814:

There are hundreds of factories in Manchester which are five or six storeys high. At the side of each factory there is a great chimney which belches forth black smoke and indicates the presence of powerful steam engines. The smoke from the chimneys forms a great cloud which can be seen for miles around the town. The houses have become black on account of the smoke … To save wages mule jennies have actually been built so that no less than 600 spindles can be operated by one adult and two children. Two mules, each with 300 spindles, face each other. The carriages of these machines are moved in one direction by steam and in the other direction by hand. This is done by an adult worker who stands between two mules. Broken threads are repaired by children (piecers) who stand either side of the mules … In the large spinning mills machines of different kinds stand in rows like regiments in an army.

As Sale writes, this was an alteration 'to dwarf even the considerable upheavals' of the previous centuries, including the enclosure of land which brought the loss of commons and enforced urbanisation. In the first issue of Dark Mountain, Simon Fairlie demonstrated that a significant part of the population either made a good living directly from the commons or depended on them to meet their needs when times were hard. If you had free and open access to food, grazing pasture and wood for shelter and fuel, then you did not have to be at the constant beck and call of farmers, proprietors and landowners. Indeed, many labourers and artisans worked only as long as was needed to ensure that the immediate needs of their families were met; the idea of working to the clock for extra income would have seemed somewhat ludicrous. Time was not yet seen as a commodity – and in the complaints of would-be employers, there is plenty of evidence of the way that this degree of autonomy and self-sufficiency limited the possibilities for exploiting workers.

It wasn't just family ties that were closer thanks to pre-capitalist production methods, community life benefited as well. Of the weaving communities, E. P. Thompson writes:

> In one sense these communities were certainly 'backward' – they clung with equal tenacity to their dialect traditions and regional customs and to gross medical ignorance and superstitions. But the closer we look at their way of life, the more inadequate simple notions of economic progress and 'backwardness' appear. Moreover there was certainly a leaven amongst the northern weavers of self-educated and articulate men of considerable attainments. Every weaving district had its weaver-poets, biologists, mathematicians, musicians, geologists, botanists ... [T]here are accounts of weavers in isolated villages who taught themselves geometry by chalking on their flagstones, and who were eager to discuss the differential calculus. In some kinds of plain work with strong yarn a book could actually be propped on the loom and read at work.

Robbed of their traditional land-rights, the newly dominant mercantile classes were seeking to take full advantage of the increased reliance of the artisans and peasantry on the wage-labour system. To see the Luddite moment clearly, we need to remember that the mythologies which shape the attitudes and actions of a culture – including, in our case, the belief in wage-labour, the work ethic, proprietary ownership, profit and progress – had to be invented, developed, endorsed and enforced. The patterns of life we now consider normal could only come to dominate at the expense of traditional beliefs. And so the industrial revolution brought with it new attitudes towards work which would prove as devastating to communal life as enclosure itself.

By January 1812, the area north of Nottingham was awash with troops sent from around the country. Armed deployments on this scale were unprecedented and the local population lived in fear. The Nottingham Annual Register of 1812 records this: 'It is impossible to convey a proper idea of the state of the public mind in this town during ... the constant parading of the military in the night, and their movements in various directions both night and day, giving us the appearance of a state of warfare.' As so often, when forced to choose between trade and people, the government treated the Luddites as a direct enemy of the state (what Margaret Thatcher would have described as 'the enemy within'). On 14.th February, the Tory government introduced a bill to make loom-breaking punishable by death. When the bill was read in the House of Lords, it was received with one of the most eloquent and impassioned speeches in British parliamentary history; understandable, given that the man delivering it was Lord Byron.

Byron's speech was loaded with sarcastic references as to the 'benefits' of progress. He questioned the inferior quality of the items produced through automation when compared to similar items produced by the hands of arti-sans. He argued that the men in question had never had a fair hearing from the government (three petitions from some 80,000 weavers had been delivered to parliament in the build-up to the events of 1811), and that the area could have easily been restored to 'tranquillity ... had proper meeting been held in the earlier stages of the riots.' Instead, 'your Dragoons and executioners must be let loose against your fellow citizens ... Can you commit a whole country to their own prisons? Will you erect a gibbet in every field and hang up men like scarecrows? ... Are these remedies for a starving and desperate populace?'

Capital and the state, those two inseparable aspects of what William Cob-bett called 'The Thing', are driven by profit and power, and impervious to arguments based on knowledgeable reason or impassioned intuition. And so, somewhat predictably, the bill was passed into law.

Those arrested in November and December were exempt from the death penalty, but were instead transported to Australia. One of the men singled out as a 'ringleader', 22 year old William Carnell, was described as having 'the merit of protecting the occupier of the House, an old man of 70 from any personal violence' – in contrast to Turpin's pistol-whipping exploits.

The strong military presence and the threat of judicial murder appeared to have an immediate effect on the local population, with only 30 frames being broken in February compared to over 300 in January. From this point, frame-breaking in Nottinghamshire became less frequent, although 1812 saw regular food riots in the area, themselves a byproduct of the hardship created by the introduction of mechanical looms. But this was by no means the end of General Ludd's war.

On 12th January 1812, a finishing machine was destroyed in Leeds, and on 15th January, the city's magistrates raided a meeting of men with 'blackened faces'. Then, on the morning of Sunday 19th January, the finishing mill of Oates, Woods and Smithson, just north of Leeds, was found to be ablaze. Yorkshire, it seemed, was in the thrall of King Ludd. On 9th February, he extended his reign, when the Manchester warehouse of the textile manufacturing firm Haigh, Marshal & Co. was set on fire, destroying the machine-manufactured cloth stored inside.

When we consider the hell that the industrial revolution had unleashed, it is small wonder that the thoughts and actions of the Luddites found fertile ground in the smoke-blackened streets of these northern cities. Having borne the brunt of industrialisation for three decades, they knew better than most what factories and machines could do to the welfare of the local population. In living memory, the greater part of a well-fed peasantry and relatively prosperous artisans had been reduced to a powerless, starving proletariat – and all in the name of progress.

As the Luddite influence moved north, it became clear that far from being reactionary, the Luddites were an insurrectionary movement. In Yorkshire and Lancashire they became more highly disciplined and more overtly political than before. Here, for the first time, there is evidence of oath-taking; oaths were considered so serious that an echo of the practice can still be found in the modern judicial process, where witnesses swear on the Bible. Under British law, the act of oath-taking was punishable by transportation, no matter which cause was being pledged or who was involved. In Luddite circles, taking an oath was known as 'twisting-in', a reference to the twisting of separate threads to form a single, stronger yarn. This sworn bond was strengthened by military-style, night-time drills, with reports of 'midnight drills', 'the mysterious tramp of feet' and 'mysterious shots in the moors' reaching the House of Lords committee which carried out an investigation into Luddism later that year. As the focus shifted to larger factories, the scale of the Luddite activities led to a ratcheting of the levels of violence.

On the night of 12th April 1812, an armed band of over a hundred Yorkshiremen made their way to Rawfolds Mill, a factory owned by the hated William Cartwright. Inside the mill stood 50 steam-powered finishing machines which had put at least 200 croppers out of work. Also inside was Cartwright himself, along with four armed workers and five soldiers from the Cumberland militia. The factory was built like a fortress, with an ingenious system of pulleys and flagstones that allowed marksmen to take aim from the second floor whilst remaining concealed and protected from gunfire themselves.

As they arrived at the factory, several men came forward and used hatch-ets and blacksmith's Enoch hammers to break down the outer gates, which fell 'with a fearful crash, like the felling of great trees.' Spurred on by this, the men rushed forward and began to smash the factory windows. Then the gunfire started. Despite one Cumberland militiaman refusing to fire 'because I might hit one of my brothers' – for which Cartwright would have him publicly flogged outside the mill on 21st April – volley after volley was fired into the assembled crowd. Undeterred, the men took their hammers to the inner doors; but tight metal studs deeply embedded in the timbers made progress painfully slow. All the while, an alarm bell rang out from the rooftop. A cavalry brigade was stationed at nearby Huddersfield and the men must have known that time was against them, but still they hammered at those doors.

John Booth, a saddler's apprentice and clergyman's son, was first to be shot, his leg shattered by a musket ball. A blacksmith named Jonathan Dean was then wounded in the hand as he wielded his hammer. Knowing the game was up, the Luddites began to retreat, but as they withdrew Samuel Hartley, a 24 year old cropper, was hit in the chest. The men had little choice but to leave the wounded Booth and Hartley where they lay, if they were to have a hope of avoiding capture themselves.

Both men were still alive when Cartwright emerged from the mill, but he refused them aid until they gave up the names of their comrades. They refused. Booth died at six in the morning; Hartley survived until the next day. Others must have received mortal wounds that night, for a local min-ister, Reverend Patrick Brontë (father to the great novelist sisters) records that two days after the event he came across a group of known Luddites burying two corpses in the corner of his churchyard.

Following the failed attack on Cartwright's mill, and with the despised West Riding magistrate Joseph Radcliffe (famed for ordering poor and orphaned children as young as seven to work in the local factories) 'scouring the district for Luds', the Huddersfield Luddites began to target smaller, less well-defended premises. But the large manufacturers were still regarded as the real enemy, so a new tactic came into being.

On 18th April, an attempt was made on the life of William Cartwright. Shots were fired, but he was unhurt. Ten days later William Horsfall, owner of Ottiwells Mills, who famously declared that he 'would ride up to his sad-dle girths in Luddite Blood', was ambushed by four men as he rode home. Horsfall was shot in the thigh and died later that night of his wounds. It is widely accepted that this murder marked a turning point for the Luddite rebellion; support for their cause, previously widespread, began to dwindle in light of this murder. George Mellor (a 24-year-old cropper who would also be cited as one of the ringleaders in the attack on Cartwright's mill), William Thorpe and Thomas Smith would hang for the murder of William Horsfall on Friday 8th January, 1813.

The first Luddite executions, however, took place in June 1812, over the Pennines in Manchester. Indeed, Lancashire and Cheshire saw the greatest loss of life during the whole rising. At the end of April, a series of food riots erupted in Manchester and the surrounding towns of Bolton, Rochdale, Oldham and Ashton. Similar riots, and some loom-breaking, took place in the north Derbyshire village of Tintwistle and the Cheshire village of Gee Cross – where one man, later deported for his actions, wore a paper hat bearing the words 'General Ludd'.

The unrest came to a head on 20th April in the town of Middleton, 10 miles north of Manchester. A crowd gathered in Wood Street, at the steam-powered calico printing factory owned by Daniel Burton & Sons. The crowd threw stones at the factory windows, and in response a volley of shots was fired from somewhere within the factory. Somebody in the crowd said that they were only firing blanks, and the stoning continued. But the rounds were live, and four men, Daniel Knott, Joseph Jackson, George Albinson and John Siddall, were shot dead.

The next day the enraged crowd returned, bent on avenging the blood of their fellows, to find a troop of Cumberland militia posted outside the factory. Instead of attacking it directly, they responded by ransacking the cottages of men who worked at the factory, and burnt Emmanuel Burton's stately mansion to the ground. They intended to do the same to Daniel Burton's property, but cavalry troops turned up and started firing into the crowd. At least four men died that day, but there were also several reports of bodies being found in the woods in the following weeks.

On 24th May, 58 defendants were brought to trial in Lancaster in association with the food riots, loom-breaking and attacks on property that had taken place throughout April. The trial resulted in eight hangings, 17 transportations and 13 imprisonments. On the same day a 'Special Commission' was held in Chester to deal with more of the rioters and Luddites. Here 15 men were condemned to death, eight transported and five imprisoned. On 12th June, eight of the Luddites convicted in Lancaster were hanged in Manchester. None were repentant. Three days later, two Luddites convicted in Chester marched to their place of execution, 'followed by an immense crowd of people'.

The 'Luddite triangle' was now flooded with troops; some 6,900 in Lancashire and Cheshire as well as 4,000 in Yorkshire. The magistrates had dozens of spies working in the areas known to be sympathetic to the cause, and the government had sent a clear message that Luddism was now punishable by death. In the face of all this, the Yorkshire Luddites began to raid any properties which were known to store arms, for the Luddites were amassing weaponry. Was the insurrection about to turn into full scale revolution?

We shall never know. The government had seen the benefits of its tactics in Lancashire and was determined to repeat this success in Yorkshire. By December 1812, on the evidence of paid spies and some very questionable witnesses, 64 men had been arrested and held for trial at York Castle. Of these 64, only seven would be acquitted. Fourteen men were transported and 26 were imprisoned. The aforementioned Mellor, Thorpe and Smith were executed on 8th January 1813 and 14 more men – with an average age of 25 – were sentenced to die at York castle on 16th January.

Of their execution it has been written:

> The criminal records of Yorkshire do not, perhaps, afford an instance of so many victims having been offered, in one day, to the injured laws of the country. The scene was inexpressibly awful, and the large body of soldiers, both horse and foot, who guarded the approach to the castle, and were planted in front of the fatal tree, gave the scene a peculiar degree of horror.

That was the end of the Luddite rising. There was still some recorded Luddite activity in the months and years following the York trial, but with a marked change in motives. Prior to 1813, the Luddites had fought to save an autonomous, communal way of life based on self-sufficiency and skilled craftsmanship. Later loom-breaking incidents were almost exclusively centred around disputes regarding levels of pay.

This marks the beginning of a shift in forms of resistance. Put simply, pre-modern resistance was a fight against enclosure: a battle to save independent, self-sufficient ways of life from destruction and to prevent the industrial machine from enslaving the people. Modern industrial unrest was a battle waged after this war had been lost. Now, the focus was on justice for the proletarian victims of the Industrial Revolution: better wages, better living conditions, the right for factory workers to form unions, the right to be looked after by a beneficent state.

Despite the wars and revolutions that cost millions of human lives, every dominant ideology of the 20th century had at its heart the powerful Western industrial mythologies of progress, the work ethic, profit and growth. In Europe, the only revolution to offer any hope of something different was that of Spain in 1936. The Spanish peasantry of the time still lived the kind of autonomous, self-sufficient lives that the northern English weavers had enjoyed until the end of the 17th century. They were, to put it bluntly, far less domesticated than their counterparts in the urbanised, industrialised proletariat. So, when anarchist-inspired ideas of a free society – free that is from hierarchy, economic inequality and exploitation – were introduced to their communities, they were largely welcomed because they were already being practised.

Today, in the West, the ideas that the Luddites fought for look prehistoric. Our societies have been remade in the image of capital so that it is hard to talk about concepts like self-sufficiency, independence and the land without being immediately dismissed by progressives on right and left as Romantics – not to mention, 'Luddites'. Elsewhere, these ideas still have some purchase. Arguably, the only viable alternatives to the dominant progressive ideology are peasant-based movements like Brazil's Movimento sem Terra (MST), Mexico's EZLN (Zapatistas), South Africa's Landless People's Movement or India's Bhumi Uchhed Pratirodh Committee. They question the standard mythology of an increasingly global civilisation and offer something different to the usual progressive rhetoric. The land-based movements of the 21st century may have little hope of becoming a worldwide revolution – at least within the time-scale dictated by catastrophic climate change or peak oil – but these communities may yet prove the most resilient in the face of an unfolding collapse.

In countries like Britain and the United States, it may be too late to emulate the Luddites, but we can take some lessons from the rebellion of the weavers and apply them to our times.

This should start with accepting that the Luddites were right. Cobbett's Thing, the state-industrial nexus, is now the dominant force in the world and its mythology shapes the times we live in. Today, the mill owners are global brands. According to the stories they tell us, this should mean that we all benefit from the bounty their markets give us. But a global perspective makes it clear that this is far from the case, and even within the industrialised world, we can ask how free and satisfied we are within our enclosure.

And all of this is before we come to the ecocide unleashed on the non-human world by the industrial machine.

The Luddites did not see all of this coming, but they understood all too well the consequences for their families and their way of life. Their story offers us a realistic assessment of the powerful, perhaps unstoppable nature of the global industrial machine, but also an understanding of the role of technology in our lives. For despite the current use of the word, the Luddites were not motivated by a mindless rejection of new technology.

Reflex resistance to technology and its mindless embrace are two sides of the same coin, neither especially helpful. I actually have a lot of time for Anissimov's Transhumanists, in that I share their belief that our species can be made 'better by design'. What I don't share is their assumption about what this means. Improving the human lot through the use of technology is not going to be achieved through a combination of surveillance cameras and warheads, nor through nuclear power stations and carbon capture, nor a Singularity in which those who can afford to pay for it become immortal semi-robots.

Instead, it's going to mean developing and using human-scale technologies which can augment our liberty and self-sufficiency rather than enslaving us to a grid. It's going to mean handlooms rather than wide-frames; control by the people rather than control of them. The best way to avoid being controlled by technology is to be in control of the technology you use.

This is already becoming a reality. On the many websites which encourage a little technical tinkering, you'll find that a combination of free and open information, Open Source software, reduced material costs, high volumes of useful waste, and micro-innovations are making it possible to develop and create projects at home – from bicycle trailers to slow cookers to mini robots – that would have needed highly specialised multi-million-dollar factories just a few years ago. The amazing Afrigadget site chronicles stories of creative individuals who build a range of tools from next to nothing. (Frederick Msiska, a peasant farmer from Malawi built a mobile phone charger from his toilet and some leaves.)

In other words, we are approaching a position where it may be possible to create once again an infrastructure built upon localised, craft-oriented, community-based, ecologically-sensitive production techniques – in other words, to return to something like the pre-capitalist idea of the cottage industry which the Luddites fought so hard to defend. It's a world in which not only is it easier to work in and from your home, but it is easier to work away from the growth-addicted world of capitalist production. With a renaissance in traditional crafts, could the artisan yet return from the brink of extinction, even as progressive civilisation itself begins to tip over the brink?

Writing of the legacy of the appropriate technology movement of the 1970s, John Michael Greer says:

> The appropriate tech movement, with some exceptions, tended to avoid the kind of high-cost, high-profile eco-chic projects so common today. Much of it focused instead on simple technologies that could be put to work by ordinary people without six-figure incomes … Most of these technologies were evolved by basement-shop craftspeople and small nonprofits working on shoestring budgets, and ruthlessly field-tested by thousands of people who built their own versions in their backyards and wrote about the results in the letters column of Mother Earth News … The resulting toolkit was a remarkably well integrated, effective, and cost-effective set of approaches that individuals, families, and communities could use to sharply reduce their dependence on fossil fuels and the industrial system in general.

Such a toolkit is needed again, and is starting to appear, in response to the crisis of consumer civilisation. I see at least some spark of a hopeful future in the development, practice and sharing of what I call ADApT (Anticipatory Design and Appropriate Technology) initiatives, which help us to distance ourselves from the corporate leviathan and restore some of the freedom of action and creativity that the Luddites went to their graves to protect.

†

In memory of

James Haigh, Jonathan Dean, John Ogden,
Thomas Brook, John Walker, John Swallow,
John Batley, Joseph Fisher, Job Hey,
James Hey, John Hill, William Hartley,
Joseph Crowther and Nathan Hoyle.

Murdered in the name of Progress
on 16th January, 1813.

References

Warren Draper, 'The Work Aesthetic', *The Idler No.44* (2011).

Simon Fairlie, 'The tragedy of the Tragedy of the Commons', *Dark Mountain: Issue 1* (Dark Mountain Project, 2010).

John Michael Greer, *Green Wizardry* (New Society, 2013).

Eric Hobsbawm, 'The Machine Breakers', *Past & Present 1* (1952).

Kirkpatrick Sale, *Rebels Against the Future: Lessons for the Computer Age* (Quartet Books Ltd, 1996).

E.P. Thompson, *The Making of the English Working Class* (Penguin Books, 1980).

Confessions of a Neo-Luddite

TOM SMITH *in conversation with* CHELLIS GLENDINNING

Issue 8, Autumn 2015

Released in 1994, Chellis Glendinning's book, *My Name is Chellis and I'm in Recovery from Western Civilisation*, captured broad discontent with hyper-technological modernity at the cusp of the digital age. The book was a touchstone for a movement of 'neo-Luddites' throughout the '80s and '90s worried about the asymmetry of power which occurs whenever the centralising machine of civilisation meets the convivial, the rooted and the 'inefficient'.

The breadth and erudition of her thought before and since *My Name is Chellis* continues to be unsurpassed, published in such work as *When Technology Wounds* (1990), *Off the Map: An Expedition Deep into Empire and the Global Economy* (1999) and the poetry collection, *HYPER: An Electromagnetic Chapbook* (2014). After spending her adult life in the swirl of radical politics in northern California and the upland desert of New Mexico, Chellis made a leap into the unknown. She lives now in Chuquisaca, Bolivia, and that is where I found her.

TS: In your new online book, *luddite.com*, marking the 200th anniversary of the Luddite uprising, you note that 'the concerned will rave about war, poverty, oil depletion and climate upheaval – as well they should. Some venture to name racism, capitalism, empire; cruelty and greed can be high on the list. But technology's role in shaping these same tragedies handily slips from the perceptual gaze.' I guess I'd like to open by addressing this slipperiness. What, for you, is technology and why is there an imperative to bring it into the perceptual gaze?

CG: Technology is a way of life. The bold thinkers who went before us included the original Luddites reacting against the industrial revolution and such writer contemporaries as Percy Shelley, William Wordsworth *et al*; plus the second generation that included scholars Lewis Mumford and Jacques Ellul. To all of them, technology is not just a hair dryer or a late-model truck. It's a way of thinking, a way of organising society, a way of being. The very organisation of a contemporary house with its separate rooms for cooking, eating and sleeping, is based, like a machine, on fragmentation and efficiency.

But you are so right: technology as a force in the rampage against the Earth is omitted from social-change discussion. Glaring oversight, eh? I remember when Jerry Mander first started talking about technology's role in bringing about the nuclear arms race in the 1980s, few in the Left in the United States saw what he could see; they thought he was 'too far out there'. But then there were those of us who had learned to think in a systemic way from our delving in that wing of the feminist movement that looked beyond civilisation to pre-patriarchal cultures. We had learned to scratch deeper; the core of the problem wasn't just about salaries and who did the grocery shopping. In the holistic health movement we learned to think about the human body and its relationship to the world as a whole, not merely as the expression of the mind/body or human/nature split that dominates allopathic medicine. The indigenous-rights effort that was growing and reached a grand denouement in 1992, on the 500th anniversary of the arrival of Columbus, taught non-natives to question society from a land-based, spiritual perspective, while simultaneously Western philosophers were challenging the post-World War Two malaise with fresh ideas about taking the very nature of modern society to task.

TS: You assert, along with Ellul, Mumford *et al.* that modern society is rigidly technologically organised. To play devil's advocate, though, as those in what seems like religious thrall to the possibility of a Singularity would put it, there may be no decisive break. In his recent essay 'Planting Trees in the Anthropocene,' Dark Mountain's co-founder Paul Kingsnorth quotes an assertion by technophile author Kevin Kelly that 'We can see more of God in a cell phone than in a tree frog'.

CG: [laughs] The cogent question would then be: But do we see more of the Goddess? Kelly is right. God has become the narcissistic, authoritative, punishing deity removed from nature and poetry, dividing sky from Earth, intellect from sensuality, right from wrong, humanity from all the 'lesser' beings – while the Goddess has been disappeared. I mean that in the sense of Latin American dictatorships: carted off, raped, drugged and thrown out the hold of an airplane into the ocean.

And on the eighth day God made the cell phone. Inherent in its processes and qualities lie the values that God pushes. It is made of materials whose extraction is foisting the bleeding of toxic materials upon the soil, water and air of *Pachamama*. Also the murdering of indigenous peoples via civil war, as in the battle for cobalt in the Congo. This is God's work: to be so powerful that He can destroy all that exists, while rationalising that such a murderous thrust equals Progress. A tree frog, though, is a perfect creature to show the bent of the Goddess. Small. Humble. Ever so lovely in its colours. Communitarian.

But, really, aren't we beyond the polarities that the modern world has palmed off on us? As a means of conceptualising the Whole of the universe, I can look to the invention of diverse characters to embody qualities amidst our utter lack of comprehension of the mysteries of this life. When I lived in the upland desert of New Mexico, I was surrounded by *Chicanos* whose roots were in pre-Colonial Mexico, who felt strongly about Nuestra Señora de Guadalupe, a goddess of kindness, care and connection. Now, here in Bolivia – of all things – I live in the city whose patron saint is ... the Guadalupe. The Greeks gave us a laudable pantheon of goddesses and gods, and through their exploits and relations we can penetrate into the workings of the human mind. And how about the spirit energies found in the indigenous world?

TS: Your most recent work operates as a reflection from the inside of the vigorous movement of 'neo-Luddism' in the '80s and '90s. Given that the original movement of Ned Ludd's followers posed such a passionate threat that it was descended on by more soldiers than were fighting Napoleon on the Iberian peninsula at the time, do you feel contemporary (neo-) Luddism lived up to the inheritance of its name?

CG: My book-blog *luddite.com* is a memoir of the wildly astute activists and thinkers who came together to attempt to catalyse a widespread movement against the unfettered, hysterical, Gorgon-like explosion of technologies that came of unfettered, hysterical, Gorgon-like capitalism. Enough intelligent folks around the world had written books, made films, given speeches, gone on strike from as many endangering technologies as possible, and/or battered nuclear weapons casings, that it seemed high time to see what we could do.

Do I think this third generation of Luddites lived up to our inheritance? Well, Kirk Sale, Stephanie Mills and I addressed this very question in a conversation that was published in the US progressive magazine *Counterpunch*. It was called 'Three Luddites Talking'. I was feeling frustrated. It was 2009, and due to a number of dynamics beyond our control, the Jacques Ellul Society had fallen apart; some of our members had taken up other callings like anti-globalisation work and the political wing of bioregionalism, secession. Computers had come in full-force and taken over the minds, hearts and free time of whole new generations. There were so many nasty invasions into our lives that viable movements in general had become harder to launch. Citizens everywhere were feeling overwhelmed. It was hard for me to accept that the ever so cogent factor of technology had fallen from consideration as people were walking around like robots with their omnipresent telephones attached to their brains.

I have here a snippet of our conversation that I'd like to read to you. It's about how our efforts as neo-Luddites fell by the wayside:

KS (Kirkpatrick Sale): There were some great and heady moments (at Jacques Ellul), some excellent conferences, some inspiring speeches, a lot of important friendships. But it wasn't really a movement and we all knew, as Stephanie suggests, that not only were we in a distinct minority but a minority regarded by many as not quite sane.

SM (Stephanie Mills): The effort petered out ... I believe in art for art's sake and discourse for its own sake. I think intellectual conclaves are worth doing if only to gather and tone up the widely-scattered intellectuals involved. But those are expensive activities. And we were fortunate to have been participants. Now we have to maintain that perspective in our several settings, along with doing the homely work of surviving at the margins.

CG: Well, I don't think the effort 'petered out'. I'm more in a *hasta-la-victoria-siempre* mood. As long as there is oppression, there is resistance; so long as there is mass technology organising life for efficiency and aggrandisement, there are people for decent values. Humans have a deeply embedded knowing when things are wrong.

TO me, what happened to our generation of Luddites is that when the 'new technologies' took hold, they literally reconfigured the patterns of connectivity. I'm talking about computers and cellphones and BlackBerries, mega-freeways and shopping malls, the Big Boxes, genetic engineering and websites, hyper-surveillance technologies – and giant transnational corporations took over our arena of expression, the publishing business. Communities that had made their way via land line and letters and meeting in cafés disintegrated. I think for a good ten years folks like us were confused, left behind. Or we were left striving, against the grain, to catch up. Or we fell into new groupings connected by new means.

Or we simply became isolated in a world of near-total technology encasement. This new world caused some of our colleagues to forge a politic shaped by different words and concepts and – for fear of being dismissed by all the people with their laptops and iPods – to purposefully stop talking about technology's centrality to control and oppression.

KS: ... It ended because it lost. The other side won! Think of the transformation of the world in the years since, let us say, 1990. All the things Chellis mentions, fundamentally based on the computer chip, swept social and economic worlds with a tsunami power within a decade, breezed past Y2K and penetrated every profession, every setting, every means of communication, every transaction. How could any critique of technology overcome that? What sense did it make to go on saying that there will be ugly consequences, that there are terrible downsides? It was – and is – inescapable, and getting more so. Even if anyone wanted to believe it – and I think many did, or as the New Yorker said, 'there's a little bit of the Unabomber in all of us' – no one, individually or collectively, had the power to stop the technological onslaught. It was the way of life chosen by the economic and governmental powers-that-be, with all the money and all the laws, and it could not be stopped.

TS: What gives me hope, energy and nourishment are the exciting changes afoot within feminism, philosophy and other fields of what have historically been abstracted intellectual thought. Concepts which previously would have been off-limits – such as non-human agency (Jane Bennett, Andrew Pickering), embodied philosophy (George Lakoff), quantum entanglement (Karen Barad), animism (Isabelle Stengers), panpsychism (Alexander Wendt/David Skrbina), transcending the Cartesian subject-object division (Donna Haraway) – are suddenly taken more seriously, or at least placed on the agenda at the 11th hour. Much of this has been spurred on by the gradual percolation into intellectual life of insights from quantum physics and other sciences characterised by unimaginable complexity, of the outdatedness of the notion of a Newtonian universe comprised of discrete, separable bits and atoms. Metaphysics is back, and it is finally acknowledged that non-human beings have an existence and power apart from human interpretation of them. Cultural mythologies seem to be shifting in areas which formerly would've been been resistant.

CG: You give worthy words to the dare of going forward. First off, it would be important to recognise that to the polar bear clutching a throw-rug-sized piece of ice, it is already too late. To scientists who in the 1990s predicted that in ten years it would be too late, it now is.

At the same time, we humans have a remarkable ability to honour the past, to live in the present even as we think about the future. Somehow, we just keep on keepin' on, don't we? How gratifying that – whether facing death or life – the realisation that metaphysics 'is back' becomes a pathway to possibility. Not to mention a symptom of the fact that, at least in terms of those of us at ground level, the schisms perpetrated through the advance of civilisation and its manifestation in Newtonian-Cartesian thinking have for some time been engaged in a process of reunion. Biophysics. Psychohistory. Ecological biology. Holistic medicine. Interculturalism. Ecopsychology. Worldwide Buddhism. The teaching of shamanistic practices. Body-based psychotherapy. Even the Pope is welcoming gays back into the church! Such arising is cracking those rigid vaults of dysfunctional thinking – and so onward, please, as if survival matters.

TS: But suppose the debate has already been lost? Particularly when the intensity and collectivity which appeared to characterise the neo-Luddites has fizzled out, been drained by the internet age, or gone quietly away.

CG: When I saw the film *Reds* for the second time, I mused 'What if they had won? How would our lives and struggles be different today?'

I am brought back to the wisdom of Andrew Schmookler in his *Parable of the Tribes*. He points out that whenever a dominating force exists – whether it's a violent father within the secrecy of a family or a world power with genocidal weapons at its fingertips – everyone within reach is changed; everyone in some way must address that dynamic. Perhaps by shutting up. Perhaps by open revolt. Maybe by telling the truth about what is going on. Or by pretending acquiescence. Edward Said echoed that insight when he wrote that everyone on the planet has been injured by imperialism. We could amplify his words by saying that every one of us has been changed by weapons of mass destruction, by corporate globalisation, by electromagnetic-run communications … you and me … everyone.

Some say that these dominating behaviours go back to the industrial revolution, others that they date to the Neolithic, others the beginning of active hunting in the Paleolithic. Some challenge the belief that they are organic to the human species; whatever, they are what has been going on for a lively chunk of our history. I think by finishing up my own personal healing process from childhood violence, as well as by entering this rich state of elderhood, I have questioned my burning requirement demanding utopia and come to better terms with the tragedy of the Here and Now. Whether you believe the problem is ingrained in 'human nature' or is an unfortunate and grotesque wounding we have endured, the horrific condition that currently infuses and impacts us, shapes and textures our every breath, has in some way been faced by our fellow beings for a very long time. We can find strength in that realisation: for all the suffering, all the knowing that the world has gone mad and we are now assembled at the brink of collective death, *we are not alone*. Our task becomes no different from that of an Egyptian slave whose life is hauling rocks to build a pyramid, a Native dodging bullets in the 1890s, an African chained to the hull of a ship in 1750 or a US soldier forced to kill Vietnamese civilians. Our task is to heal. It is to teach beauty and preserve what is archetypal to our species, to tell the truth, to fight against what is wrong, wrong, wrong; to live in praise of life's very existence.

TS: An inspiration for this healing from South America which has always stayed vividly in my mind is recounted by Daniel Everett in *Don't Sleep There Are Snakes*. The Pirahã tribe he lives with are remarkably resistant to the wonders of technology, right to the level of agricultural techniques. If they are given a piece of advanced technology, they might use it. But when it breaks, they'll never look at it again. They are so confident in their way of life that they deem everything else inferior, unnecessary. If their house blows down, they laugh at their misfortune and build another one.

Returning to the idea of Progress, and specifically with your experiences living in Bolivia, is there anything you've learned from the relations of people there to the *techné* they engage with on a daily basis? Is there intelligent resistance happening or are they just desperately waiting for access to the wonders of modern life?

CG: That slippery notion of 'Progress' – whether conceived as a psychological, evolutionary, social or technological event – is so ingrained in the modernist mind-set that it's assumed. I mean, given the precarious disorientation and terrifying alienation we experience in a life removed from our given habitat, the natural world, wouldn't you rather think that you're on the verge of getting a more magnificent pill to cure your ailments, a more glamorous apartment, an unimaginable new lover or a more clever telephone? What could be better than Virtual Reality when the reality of this world is so forbidding? Or the assurance that any day now scientists will announce a mechanical-miracle solution to global climate change? Even Marx himself presented a view of social change based on a notion of Progress.

In the 1970s and '80s the Hopi people of Hotevilla had a raging debate: whether to accept the electricity the state insisted on install-ing. Those traditionalists against the incursion argued that it would destroy their ancient culture, while the 'Friendlies' in cahoots with the outsiders craved electric heaters, night lighting and TV. When Australian aborigines were confronted by a mining company intent on drilling into their sacred mountain, they freaked; to them the mountain was a sleeping lizard whose job was to dream their existence – and the drilling would wake the beast up. And remember the U'wa of Colombia that threatened suicide if multinational Occidental Petroleum dared to abuse their land?

When I first traveled to Bolivia in 2006, I was drawn to its clear skies, cobblestone streets, antiquated automobiles, absence of cell phones and nuclear plants, dazzling intellectual life and strength of friendship. I fell in with people who cooked on woodstoves, celebrated the traditional holidays, marched for justice – and read Kropotkin, Sachs and Klein.

All that has changed. The jump from countryside peasanthood and early modernism in the cities to globalised spectacle is causing disori-entation of the kind that leaves its citizens stunned. The government is centralised and authoritative: whatever Evo wants, Evo gets. He wants satellites, freeways, nuclear, 4G, big industry and smart buildings. I wouldn't say that the majority of Bolivians were champing at the bit for internet and iron mines, but then as these became the chosen conduit to trickle-down cash, many have fallen in step. The strongest resis-tance comes from the gatherer-hunter peoples protesting a high-tech highway through their constitutionally-protected nature reserve, and they have truly captured the hearts of Bolivians.

TS: So you want to save what's left?

CG: I came to Bolivia to be part of history. Tom Hayden had invited me to travel here in 2006 for the inauguration of the country's first indige-nous president. After all the dictatorships and US-sponsored military governments, the people were literally dancing in the streets. Everyone was talking politics. And hope. At the time we in the United States were saddled with George W. Bush, and let me tell you: it was grim. How refreshing to see and feel excitement about the future!

I also decided to move here because of the level of technology. I mean, one has to feel some techno-envy for van Gogh or the poets of the 1930s in Paris, yes? Bolivia had the occasional donkey on the avenue, *campesinos* in from the campo in their traditional hand-woven ponchos, adobe *hornos*/ovens in the yard for baking bread, and home-made corn liquor *chicha*. Café life was highlighted by vibrant political discussion and artistic creativity. I wanted this in my life! And rightly speaking, I could see that one visit a year was not going to cut it; I had to live in Bolivia.

Indeed, in my five and a half years here, I have witnessed and participated in history. I could never have imagined the sequence of events that was to come. Bolivia is a country engaged in the making and remaking of anti-imperialist socialism in the global context of power-over politics. For all the impressive talk about the rights of *Pachamama/ Madre Tierra*, the lunge toward fast and furious industrialisation dominates. There are humongous new mining excavations. Oil and gas pipelines. Deforestation at a clip. Nuclear power plants in the works. Telecommunications towers spreading 4G radiation contamination. A telecommunications/ surveillance satellite. A cell phone in every hand. State-of-the-art tanks and planes for the army. High-speed freeways linking Latin American countries for participation in the global economy. Blame and violence against any who appear to oppose what's being done. Diminishment of freedom of the press. It has all happened so fast, infiltrating in the inevitable silence of 'Progress.' And in the top-down claim to power of authoritarian government.

A bizarre historical theatre is unfolding here in the *altiplano*, and Bolivia is not apart from what is happening in every corner of this *pobrecito planeta*. At the same time, it's a place of lavish surrealism. Of feral traditional celebrations and crashing military marching bands. Of scandalous poverty and epic national pride. Of explosive rebellion and tightening social control. Of deep intelligence and ludicrous errors.

I'm a fan of spontaneous creative expression, and here in Bolivia you never know what's going to happen next. All the world's a stage, you know. I deem that, in these precarious times that are so underscored by conspicuous insanity, we might consider cultivating such an outlook.

Prospecting for Equanimity

JASON BENTON

Issue 7, Spring 2015

> The stakes are in the meadow ... the fields are overgrown
> The winds of change are blowin' through the place I've called home
> They're digging at the edges to build the power line
> Same old story ... but now the story's mine ...
> It all began 300 years before
> What story is beginning
> If this one is no more?
>
>   – Railroad Earth, *Lone Croft Farewell*

This story is nothing special, nothing new. It is a story of time, my home, my family, and how this story continues to consume our time and space to be human, creating existential crises in individuals (myself). Environmental writers write about childhood because, it seems, this is the only place where, if lucky, we are now afforded the habitat to be human for a short while before we're forced into the fold. This is not one of those articles. It is a recounting, in my words and my grandparents', of how things have changed and of the unfolding discontentment across generations. Out of this arises a patient hope – for what, I still cannot say – but it is not for what we have now.

This place, Western Pennsylvania, the western foothills of Appalachia, has produced a lot of coal, gas, and steel, (and a lot of broken backs, broken families and cancer) but also Annie Dillard, Edward Abbey, and Rachel Carson, though most of us around here have never heard of them. What abides are strong family ties and proud gardeners, a connection to the land (what is left of it), and a stubborn independence, although for how long I do not know. It is a place I will continue to call home until I die or it is taken from me.

I left here as an eager teenager and returned a decade later married with kids and a mortgage on a four-acre plot. A curious thing happens after spending time away. Aesthetics and idiosyncrasies of time and space – a workday commute, a weed, or ambient sound – become more pronounced, as the difference between the present and memories of the past reveals the quality of relationships, change, and often injustice. Discontent and love arise hand-in-hand as objects and experiences resurrect these memories, revealing something new and something old and a living story which lies in between.

Some of these objects and experiences are products of civilised life. Civilisation goes about its business condensing the raw elements of life (including people) into capital and products, and in the process condenses the experience of time and space into things like gasoline, trucks, p-values, and phones – potent mediators to the experience of life and reality. This business permeates even my thoughts, behaviours, and sometimes desires. In the process, stories fragment and memories fade. It's the 'same old story' once again.

But there is always room to rewrite, recreate, and be reborn. I'm still just as eager for something different, something new – perhaps because I was born of this place which has been in constant flux since my farmer ancestors immigrated here from Germany, Ireland, Scotland, and Denmark over two hundred years ago.

This eagerness is not for what most would think. I've travelled enough of the world to find that the best things in life come from my garden, kitchen, and bed. And an adventure away from it all, a solitary place to get lost and lose myself, I can often find within my garden or the local woods. I spent most of my time away living in Boston. I do miss its diverse and witty people, the White Mountains nearby, and a walk to the Atlantic for a dip. But I wouldn't call it home. *This* will be – probably until I die. This will be the place where I'll always return after getting lost, after losing myself. After many generations of being pushed aside, something within me has decided that it is time to remain still and quiet, at least for now.

Much of this place has changed. The fields are now dotted with giant houses and gas wells. Gas lines vein and artery the land, ever-vascularising new territory and depths, varicosing to the surface for stations, valves, and utilities. Churches have water ministries (not to be confused with baptism) helping families with contaminated or lost water. People seem richer and poorer, which is nothing new to this business. Once non-existent or seldom-seen, smart phones, stink bugs and drugs are now ubiquitous – luxury and its victims reinvented.[1]

The woods are posted. When I was little I used to be able walk miles in most directions from home without bothering anybody except Mom who would, having exhausted her voice, wail on her hefty iron triangle bell to call us in after we were gone too long. Impossible now without breaking some law, namely trespassing, as the countryside is increasingly owned by businesses or individuals who live far away. I remember her heaving the antique from its nail on the wall, a symbol of a time when kin worked within earshot – although in her case she slept alone every night; still does, with Dad working night shift at the plant.

We boys roamed the countryside, pretending to be trappers and home-steaders like Daniel Boone, but found someone had already beat us to it and then disappeared, leaving behind ploughs, animal traps and trash. Those relics became our rusty treasure, shifting our ambitions from rocks, plants, and animals to more neurotic, civilised pursuits: archaeology, anthropology, and something else – for those rusty relics became something like Yorick's skull, leading to various questions of existence:

'Who were they? Where did they come from? Where did they go? Where did I come from? Where am I going? Who am I? And if all the Daniel Boones are gone – their tools, land, and dreams absconded – where is this fantasy taking me?'

When I was 12 my brother, a year younger, wept when we learned that someone was planning to build in the woods behind our house where the old sawmill used to be – where we played like giants along *the* stream, shot, trapped, and caught countless helpless critters. We booby-trapped the place with animal skulls and snakes to dissuade them – it worked, so we thought. That place was sacred to us. We still pause each time we pass there, now with our little ones, in awe of how tall the sycamores have grown, how little we now feel. We stand mouths-agape peering at the ivory heights, wondering if sycamore fruit is edible, eyeing a viable avenue to the top.

'I wonder what those taste like … Hey, where are the kids?'

When I was a child, time was absent, but now everything is moving and accelerating, especially those little ones. My great-grandparents' world was no less chaotic, but for several generations prior to theirs there was a relative homeostasis, a sameness from one time to the next (they would have disagreed). Change has been constant, bulldozing the past, yet still dictated by the same intensifying powers – powers that as I grow older I find all around me and even within my own conscience; and sometimes, to my annoyance, my unconsciousness.

I try to pluck such infiltrates from my soul just as I do the stink bugs (an Asian insect who stole away on Chinese imports during my decade away) from every nook and cranny of my home, or as I once plucked them from my toddler's shoulder. As for my children, I just can't say what time will be or bring for them. Will they stay? Will they return? How will the pressures of time and these powers form them?

The mining frenzy hasn't changed, even since the first Drake oil well was drilled just north of here in 1859. We're always moving and being moved by the energy industry, which powers what William Cobbett[2] called 'The Thing', and others have called progress, GDP, revenue, industrial warfare ('the War effort'), democratic capitalism, The Machine, The Man, development, growth. Grandpap recently shook his head in disgust – shocked that they all-at-once strip-mined *and* fracked the vast rolling fields across from his childhood home:

'That was still good land! Been farmed long before I was around – a lot of people took care of those fields.'

The Thing has given us much, though. Compared to my great grandparents we live like royalty – more convenience and luxury; however, less freedom and less time. If they saw us now they would quietly nod and politely say, 'That's nice,' then gratefully, skillfully return to their work. I romanticise, but these were folks who lived life abundantly, and without electricity or indoor plumbing. My brothers and I have careers now, and if you've got one around here you are likely tied to mining or the energy industry, and invariably tied (or perhaps noosed, knotted, lassoed, or wedded) to The Thing.

One of us counts and recounts the surplus, one ensures it is safely acquired, and I manage the side effects – as a psychiatric nurse. I say *manage*, for true prevention would be anathema to growth, to luxury – to the noose. At work I see the double-edge of our way of life: the rich man who feels alone and empty, and the poor single mother who is exhausted from working three jobs, neither of whom have time for their children. Cobbett protested against this onslaught, for fairness, longing for the nation of his childhood. But turning back the clock is futile, and with only three weeks of holiday a year, there's no time left for dealing with 'hucksters, governments, or favourable representatives.'[3] The soul needs daily restoration: at home I have my children's youth and my wife's love to enjoy, there are beans and berries to pick and preserve, someone needs help with a shovel, the winter wren offers a song and dance, there's food to cook and share.

Dozers and trucks, symbolic and steel, can become deeply personal. Grandma had the noisy, dusty coal processing plant nearby, their dairy farm razed for progress, which is now Moraine State Park. My childhood home (the actual house) had been relocated from its hilltop farm to the edge of the woods in order to make way for strip mines. I have peered and peed into more than one open mine pit. Stepping from a wet vernal paradise teeming with life to a silent dusty pit the size of a small village, I found myself asking then, as I do now, 'Is all this really necessary?'

My friend lived in the trailer park down the road – his father was a dragline crane mechanic, always reeking of diesel, piss, tobacco and sweat from sleeping in a rig or his pickup. They would operate those draglines 24/7: one was a mile away, lighting the horizon and drumming the earth. You didn't notice its constant racket until it suddenly broke down, sometimes with a thunderous echoing boom as the house-sized bucket fell to the ground. Right then, an etheric, proverbial breath was released: silence wind birds breath heart and mind quickly settled to the fore of consciousness. The land and I exhaled, and for a brief moment possibility opened her door.

Equanimity restored.

For as long as I can recall I have been awakened by the drone of early-morning coal trucks on nearby Routes 19, I 79, and 422. As a teen I spent the night at my girlfriend's (now wife's) parents' house. The road lay parallel, just a few yards away, to a couch in the foyer where I slept – once. Jake-brakers! The noise of slowing trucks is loud enough to shake houses, make dogs bark, babies cry, throw china from cupboard shelves, and unhinge windows and your mind.

Today the morning drone is accompanied by cavalcades of frack water and rigging trucks. On my morning commute I share the highway with the trucks carrying coal, water, pipes, and rigging, and also windmill wings hundreds of feet long on their way to the Appalachian mountaintops east of here. There are pillars of fire on every horizon, which always lend images of a biblical Israelite camp and tabernacle, like the posters on a Sunday school wall.

At work, many of my patients are coal miners, roughneckers, truck drivers, and their wives and children – the children I worry for the most (and what of their children's children, including my own?). Too many of those young people look at their exhausted, indebted, irritable, unhealthy, addicted, and divorced forebears and ask me the question: 'What's the point?' without a story to tell or fulfill. I have no answer that will satisfy their soul and I almost appreciate their protest: sitting idly, already disillusioned, without ambition or hope.

The miners come with bent backs and spirits, fearing they are no longer useful. The roughneckers always find a way to gaol: heroin, fights, more heroin. The truckers, anxious and paranoid: running against time, traffic, and debt. There are also surveyors, land grabbers, rich well owners, poor pipe welders, and an arrogant rookie accountant setting up some kind of hedge fund for the drilling companies. Many, though, are jobless, clinically poor, and by the time they see me are beyond their wits' end with their struggle. People around here (myself included) aren't apt to ask for help, particularly from a stranger, and more particularly from a shrink. During our lunch conversation my co-worker tearfully shares the news: 'It's a malignancy'. She's the fourth in our office diagnosed with cancer this year. My boss complains that 'They're dumping in the stream again' behind her house. She isn't sure how to stop them or clean it up.

The daily 10am and 2pm trains rumble and screech outside my office window, a mile of coal, oil and chemicals in tow. I drive home with the trucks and need to use my windshield wipers when following the frack water salvage trucks as they leak an oily, briny residue onto my car (destined be dumped in old wells in Ohio). On the car radio, I listen to a politician argue that in order to save the future existence of Pennsylvania farming we need more gas operations on farmland.

I arrive home and my daughters and I wander into the garden where dull bituminous and shiny anthracite lie on the ground, dropped from previous mining operations. We pick tomatoes, flicking off confident stink bugs, while the Haliburton pickup drives by and the Shell helicopter flies over-head – every day at 5:45pm.

Entering through the basement door, I'm greeted by my radon detector beeping: the level is still above the EPA 'safe' level, even after the expensive mitigation system whose constant hum drives me mad. Some wondered if radon had contributed to my aunt's early death since she lived in an under-ground house near gas wells and mining. We'll never know.

The calm of the evening is periodically interrupted by the wobbly fan blades of my new electric ultra-efficient heat pump, my attempt to warm and cool my house 'sustainably'. I listen with indignation and vengeful pleasure. I dare not fix it. The blade might one day let loose and destroy its innards, but that is fine. My beloved winter wren couple (and since, a host of other birds, chipmunks, squirrels, a wild kitten and a baby crow) recently nosed around inside looking for a place to build their nest when the fan kicked on. You can't kill pure spirits like theirs without damage. Winter evenings are warmed by my pellet stove: an insane contraption, always needing fixing, and relying on wood for sawdust, volumes of natural gas to dry and press the pellets, oil to haul and package the pellets, and coal-fired electric to power the thing – all for a simple warm fire. We're burning even the dust and selling it as an 'ecological' heat source, referring to sawdust as waste. Not even the dust is allowed to fall as an offering to future forests, fungi, termites, and dirt.

But The Thing is there, so I use it. I'm as patient as a snake, but sooner than later, in a maniacal fit of rage and flurry of pellets, the Thing will be speeding to the scrap yard.

'Mommy, where'd Daddy go?'

'He'll be back, babe. Let's clean up these pellets.'

I later turn out the lights, burned at the local coal (soon to be gas) plant, just west of here, which I'm told may have contributed to the asthma that nearly killed me when I was three and now suffocates my toddler. I fall asleep to the drone of the trucks.

The change since my great-grandparents' time could have been read on the painted metal signs littering my backyard when my wife and I bought this place. The signs had been repurposed by the noble man who built my house to patch the barn roof and outbuildings. But the wind, an omen, yanked them back off, sending them throughout the yard. Earth and her gravity, nature's ultimate force of fusion, began to rust, rot, and absorb them.

I've since re-repurposed the signs. They read: 'BLASTING ZONE: KEEP OUT' with a blasting schedule and siren signals; several 'Real Estate', 'Development', and 'Land' signs with contact numbers; 'Firewood, U Cut, $10/load'; and 'Church Services Parking *Only*'.

My home is just across a moraine valley from my great grandparents' farm. It's about five or so miles as the crow flies. I never knew their farm as a farm but as 'Moraine'. A few cornerstones to their house and barn now lie next to the playground above the North Shore swimming area where we swim, picnic and swing with our daughters. Grandma tells me that their ancestors' homesteads were higher up the hill above theirs. All the farms in that area (about 16,000 acres) were evacuated by the state government during the '60s, razed to build a park and conduct an experiment: to see what happens when a heavily farmed and mined land is left to grow wild again. There is even a 'propagation area' where all human contact is prohibited, although every hiker and hunter I meet in the woods has had the curiosity to explore it at least once. It was perfect timing. The mines needed a little more room (they always will) and most family farms were folding (literally, by strip-mines), and still are, everybody getting jobs in the mills and mines.

The area is now Moraine State Park: Muddy Creek is dammed, the valley filled with Lake Arthur, the fields grown to scrappy woods. Grandpap said that when he was 'knee-high to a grasshopper' there were few trees. 'You could see a long ways, so you could, and all the squirrels were gathered in stands here and there.'

We are heavily wooded now, the deer are thriving, and we have more bear and coyote than ever. The squirrels are fine for sure. You can stroll through the park's woods and find crumbs of history: old roads, fence lines, barn skeletons, stone silos, a railroad, and vacated coal mines and oil rigs, some of which might be reopened and fracked horizontally.

I enjoy walking backwards in time along electric, barbed wire, wood, and stone fences leading into more rugged territory, hospitable to nimble horse and hand farming but not to heavy machinery. A split-rail fence disappears into a steep cliff of shale tailings, a mine cart rail exits. A stone fence with a section gone, bulldozer tracks intersecting. A six-storey red oak erected through an iron tractor wheel, absorbing the spokes as though the object were never there. Stone house foundations full of broken preserve jars, pickling crocks and galvanised buckets. A concaving Victorian house with an upright piano sunk into the basement, the bench turned over with sheet music to 'The Old Rugged Cross' spilled across the wood-tiled floor. The house looking a little crucified itself, even with a crown of thorn bushes and a large maple piercing its side. I finish my walk whistling and humming the hymn, 'On a hill far away'. I never quite understood that song – exchanging material trophies for a spiritual crown. Treasure for treasure just the same.

Since the lake was 'put in' my family has enjoyed fishing, hunting, and playing in the park. But more than the land has changed since my great-grandparents lived here. After they were evicted – 'kicked out' as my grandparents say – they tried relocating to a new farm 20 or so miles to the north, but it just wasn't the same and farming was no longer sustaining or sustainable anyway. They lost all their investments, land, and neighbours. Everything was more expensive – a pig and a handshake no longer purchased things like an ornate oak dry sink. Furthermore, my great-grandfather had been weakened by the black lung and cancer he acquired from working in the mines over the hill from their place. Their family, including many cousins and in-laws, lived on that hillside and among neighbouring farms for several generations. My great-grandparents had to sell and retire to a rented house.

We haven't farmed since. Their son-in-law, Grandpap, worked in the mills after he was finished with the army and Grandma worked in the local general store, Cal's, until they retired. Most of the mills have gone to Mexico and China, and Cal's creaky wooden floors burned and rebuilt with a gas station pavilion blaring bright lights and bad music.

Shortly after the park opened, my grandfather built a pontoon boat and they took my great-grandparents to the lake for a float – once. I'm told that my great-grandparents and many other old-timers held much resentment and grief, refusing to set foot in the park.

I don't think that I would have wanted to see them that day on the lake– weakened, retired, spent, defeated, everything gone – while everyone else motored around gleefully, dozers and machinery still crawling like ticks and fleas over their cherished hills, ripping out fences, leveling centuries-old barns and buildings. That was surely not a time for laughter or wrath; they slipped into their own soul's desert and faded away3. Great-grandfather died and soon after great-grandmother could've been found any day of the week in her Sunday best, Bible and purse in hand, walking the wrong way down the highway to church.

[Whether it be a gadget, heat source, or way of life], ... we see the secret failure of American commerce. For all its obvious successes and benefits, capitalism has failed to capture our hearts. Our souls, yes, but not our hearts. There is something ugly and mean about it which most of us can never accept.

  – Edward Abbey, *Appalachian Wilderness*

After walks in the park woods I often ask my grandparents about the landscape, the farms, the people, although their recollections are beginning to fade. I recently asked about an area between Alexander's Ridge and the 528 bridge. There is a mediaeval-looking stone silo perched high on a steep hill with some 'cubby' coal mines dug underneath. The house foundation is thick sandstone just south of the silo. Farther below, the hill is stepped by cranes and dozers.

Grandpap said, 'they were sheep farmers, been there a long time. I can't remember their name. They were related to the Alexanders and that was their original farm, so it was'. It must have been idyllic with views to the south and east across the hollow opening into the valley; a high flat plateau for pasture, gardens and apple trees. The progeny from their fallen heirloom fruit are still there – old, crippled and gnarly. Pines planted on the ridge break the west wind. The lane entrance is greeted by a sandstone pillar with an iron ring set at the top for a gate. The other gate is hidden under thorn bushes. It was strip-mined just to the north where their fields must have been. The trees are all diminutive up there with broad, bald patches of shale where just a hearty goldenrod and sumac have held on for the last half-century.

Walking along there, I looked back to find my wool hat hung tauntingly by an arched thorny finger. Multiflora rose bushes blanket the woods floor; their thorns rend clothes and skin. Grandpap says it is 'an evil plant' brought in from Europe by the Game Commission during the '40s and '50s to provide wildlife cover among the tidy fields' fence lines. Probably some WWI vet, a retired captain perhaps, reminiscing of the French countryside where ancient field and family lines ran together along elegant hedgerows. However here, everything is hastily built then trashed, making way for someone's new scheme – old ways, traditions, family lines and memories fragmenting and drowning in their wake. Nothing is allowed the time to take root, flourish, climax. Just as the plot thickens, the same old story starts again.

Now The Thing, that hideous rose, is everywhere and infests my yard, magically grafting itself into my rhododendrons at the base and sending a multiflora barbed shoot out the top. The rhodo will slowly morph into a thorny multiflora bush if you're not vigilant, cutting away all the thorny sections, especially the initial graft at the base.

Grandpap's father told him that the coal miners first arrived as immigrant families who dug cubby-hole mines 'like a fox or groundhog right into the side of the hill'. You can still find where these were dug with sink holes above them and oily orange water oozing out of their base. We learned to call this 'wizard's piss' as kids from my friend's grandfather who ploughed and harvested the fields by day and read Western novels by night. Grandpap said, 'They sent their kids in there with a lamp and a rope tied around them. They pulled out the coal with buckets and had to work fast before it flooded or caved. There was a lot of water, *good water*, in those hills, you know. They lived in shanty towns – Frog Town, Coal Town, Miners' Town, some in Isle – which aren't there any more, you know, some of those towns are where the lake is now.

'They were poor, had just their clothes and tools. Once the big machines and strip mines came in they all disappeared, went somewhere else, I guess. You see, us farmers had the land so we never went hungry. The miners would come up starving in the winter and offer a bucket of coal for a chicken or preserves, so they did. What a way to live.'

He scowled, thinking of being a without a good piece of land, a 'fruitful acre' – a man like most his age around here who would never mistake Allis Chalmers for a band or Gravely for a condiment, wouldn't be caught dead without an impeccable yard and garden, always eager to share something colourful and tasty growing out back.

'The farmer had it alright, so he did. You work hard. A good year let you set some aside. You saved. Nobody was rich but everybody had enough back then. Long as you took care of the land it returned the favour. Neighbours helped. In the winter you could take it easy some, hunt, enjoy your family. But you just can't do that now ...'

Grandpap drifted into memories while the rest of us were distracted by the Thing, a stink bug crawling across the television and Big Ben's fumble, the Steelers' second in the first half.

I am reminded of this story – those nomadic miner kids with their buckets and ropes tied around them 'just in case' with a specific knot which prevents severing the man (or child in this case) you are trying save – every time I drive by the Coal billboard on Route 422 across from the fairgrounds. I would never mindfully send my kids in there, yet every day I mindlessly burn the oil, coal, and gas resurrected by my friends and neighbours, their sons and fathers, my injured and worn patients. The billboard reads 'Our Gift: 250 years of reserves' with a little boy and girl holding hands on a green lawn. If you start with those first poor miner kids, we probably have about a hundred years to go at most. But we're fracking deeper and farther now, so I hear that should give us a hundred more at least. Maybe some time then or another hundred years after, all at once (and maybe once and for all) the door will open again.

The wheels of the world are moving on
And if the door is closing
I guess it's time I'm gone

— Railroad Earth, *Lone Croft Farewell*

Shortly after writing this a representative for a drilling company, an Exxon subsidiary, knocked on my front door in Prospect, Pennsylvania and asked for the rights to everything which lies from an inch below the surface of my land to the centre of the Earth. The cow pasture and corn fields across the lane will soon be an industrial complex, a 5-well frack site. He assured me, 'There is nothing to worry about … '. To his puzzlement, I assured him not to worry either because I no longer worry and no longer hope, at least for what he has to offer. We have all the time in the world. Come in, have something to drink and eat. Let me pick a tomato and some basil and I'll share a story. The bread is already in the oven. Unfortunately, he believed he didn't have the time.

Notes

1. Ralph Waldo Emerson frequently explored the karmic idea that for every material excess, there is a double-sided, moral/material victim or 'defect', e.g. the modern guilt over our excess, and our pitiful attempts at finding 'sustainable' methods of having more of the same story. See RWE's essay, *Compensation*.
2. See Dark Mountain, Issue 2, *The Shuttle Exchanged for the Sword* by Warren Draper. The Luddites' 'same old story' is what drove Cobbett to stoop to his political endeavours and my ancestors to seek a place in America where they could farm unbothered. That Jeffersonian dream remained a dream for almost 150 years until the story quickly caught up with them.
3. From Robinson Jeffers' *Soul's Desert*.

THREE

We believe that the roots of these crises lie in the stories we have been telling ourselves. We intend to challenge the stories which underpin our civilisation: the myth of progress, the myth of human centrality, and the myth of our separation from 'nature'. These myths are more dangerous for the fact that we have forgotten they are myths.

NICHOLAS KAHN & RICHARD SELESNICK — Extinction Cabinet — *Colour photograph* —
Issue 9, Spring 2016

'As I looked down I thought the ideal extinction museum would be one of indeterminacy and rumour, perhaps in a side-show cart built from the ruins of our ghost towns and dragged through the dying landscape below, stuffed with the memory of those things that could no longer be found...' from Truppe Fledermaus: 100 Stories of a Drowned World

Travelling Man

KIM GOLDBERG

Issue 7, Spring 2015

The Travelling Man rolled into town in his painted-up wagon pulled by two stout fir trees. Yes, you and I know that trees are sessile and cannot pull a peanut, let alone a 900-pound caravan, across the landscape — it must be some sort of trick. And indeed it was. The Travelling Man's engineers had laid a concealed metal track four inches beneath the earth along the main wagon road joining all the towns. At each end of the track, a steam-powered winch was hidden in a hen house. The engineers had worked away night after night with sound-deadening Teflon shovels while the townsfolk slept. The lips of the incision in the earth pressed themselves together like a kiss. They parted only briefly when the tree-platform's wheels slid through on the sunken rails.

The sight of a wagon pulled by fir trees filled the townsfolk with awe. Hearts quickened. Jaws gaped. They had heard rumours, but who could believe? 'Anyone who can harness the stasis of forests to get some useful work out of the damn things ... Hells bells! Let's see what else he's got on board!' (This reaction had been predicted by the Travelling Man's market research team, who had been required to sign non-disclosure agreements.)

He stopped his wagon in the middle of Main Street. The town's children raced in from all corners. Their porcelain skulls, thin as wasp nests, bobbed on C1 vertebrae like bright buds beneath their shiny skin and hair.

'Will it bite me?' asked a freckle-faced girl as she reached out to stroke the tree's craggy bark. (Trees of such magnitude had not been seen 'round these parts for quite some while.)

'Course it won't bite you, darlin'! Trees don't bite,' laughed the Travelling Man, who had donned his gleaming prosthetic smile at the outskirts of town to hide his teeth, blackened from years of betel-chewing.

The townsfolk followed their children, crowding around the caravan. Brilliantly coloured long-tailed birds were painted on the doors on one side. With a flourish and a piano chord from somewhere, the Travelling Man swung open the doors to show off his wares. 'Plenty for everyone!' he announced.

Gasps rippled through the crowd as the townsfolk marvelled at what they saw (although they understood none of it). They each held out cupped hands to receive their gift. For that was the genius part – the Travelling Man's miracles were free. The blacksmith went away with a small cough. His neighbour took home an arrhythmia. The freckle-faced girl was given a shadow no bigger than a quail egg on her brain.

Strange Children

AKSHAY AHUJA

Issue 5, Spring 2014

A story to start – about fertility and its demands.

A king meets a woman of astonishing beauty walking in the forest. He must have her; he is willing to leave everything behind. The woman tells him that she will stay with him under two conditions: he can never interfere with her or question any of her actions. The king agrees. The woman doesn't come to his kingdom; instead, he goes to the forest. There they live together in a state of rapture. The woman is soon pregnant. After she gives birth – alone, without aid – she rises with her newborn child, walks to the bank of the Ganga, and throws the baby into the river. The king watches her, horrified, but he remembers his promise and remains silent. Six more years, six more children. Each is thrown into the river on the day of its birth and drowns.

An eighth child is conceived – the woman, it seems, is inexhaustible. Soon, after another effortless labour, she rises and begins walking to the river. The king, after all of these years, has reached his limit. Perhaps, although the story does not say so, one can have one's fill even of rapture. As the child is lifted over the water, the king finally asks the woman how she can do this to her own children. He begs her to stop.

The woman turns to face him, the baby in her arms. The king has broken his promise, and she tells him so. She must leave now. She is, she explains, the goddess Ganga, the spirit of the river. She has her own reasons for killing the children (they were reincarnations of immortals, and anxious to return to heaven). She will, she tells the king, spare the last child, but after raising him past his infancy, she will no longer be a presence in their lives. And she does go – she disappears.

This story appears early in the narrative of the *Mahabharata*, one of India's two national epics (the earlier and more straightforward is the Ramayana). There is no definitive text of the *Mahabharata*, which has been retold with variations in every Indian regional language – by Kumaravyasa, for example, in my mother's Kannada – and is still being narrated by storytellers around the country. The epic's stories are a living part of India's culture in a way that has few equivalents in the West, especially now that the Bible is less known.

When I was growing up in Delhi, I absorbed them through everything from television shows to classical dance to my family's conversations, where the actions of the epic's characters – Duryodhana, Yudhisthira, Gandhari – are a rich source of allusion, and still used to judge the rightness of one's actions (or, more often, those of others). When I left India as a child, these stories began to retreat in my consciousness. It was only recently, as an adult, that they began speaking to me again, telling an old story of a civilisation's increasingly destructive relationship with itself and the world that sustains it.

I re-read my old comic books, C. Rajagopalachari's abridged English version of the epic, and parts of Kisari Mohan Ganguli's 19th-century translation of the Sanskrit original, which runs to over 3,000 pages (I can only read English with any fluency). I filled up with the epic again, and for the first time sensed a cycle set in motion by the story I have just told. The *Mahabharata* cannot be effectively summarised, but in broad outline it is the story of a great battle between two sides of a family which grows to include most of the neighbouring tribes. The families descend from two brothers, Pandu and Dhritarashtra, and the battle is fought by armies led by their descendants: the Pandavas (Pandu's five sons with his two wives, Kunti and Madri) and the Kauravas (the sons that Dhritashtra has with his wife Gandhari).

As with the Iliad, the somewhat more righteous side wins, but only through treachery and at immense cost. It is the story of a victory without savour, one that leads not to a golden age but a dark one – the Kali Yuga – that in Hindu thinking continues to this day. After the battle, the Yadavas, the tribe of Krishna (the reincarnation of the god Vishnu who is at the centre of the epic), which fought on both sides of the battle, will annihilate itself in internecine fighting, and the kings who won the war will hand over their power and wander the wild places of the kingdom until their deaths.

In my imagination, the epic is surrounded by waves: the narrative begins with the king's question to the river, and ends with one of the civilisation's great cities under water. Outside, it seems to me, in the ocean encircling these stories, is the world outside of narrative, or at least the type of narrative that we are accustomed to – the stories of conquest and defeat, rise and fall. This is the world the king is living in when he refuses to question to goddess, even as she throws one child after another into the river.

Somewhere, though, among the cultures that live out in the waves, a question arises. People wonder why certain children die and others do not.

They begin to feel a sense of separateness from the rest of life and each other, along with a growing conviction of what justice should look like for human beings. The king breaks his promise to the goddess. He questions her, and he will get his answer, because the goddess is susceptible to questioning, but then she will always leave.

When she leaves, a very specific cycle begins. Humans lose access to the fertility once provided by the gods, and are left with a world that is much easier for them to manipulate. The modest but fairly reliable harvest – which was always recognised as a gift, and was taken away from time to time as a reminder of the fact – is replaced by a fertility controlled by people. Suddenly the harvests – of food, animals, metals, and children – become more plentiful, and an apparent golden age begins.

The human generations created in this way, however, prove to be not quite whole. Through their origins in a divided and manipulated fertility, there is a seam inside them, one that gives them less and less understanding of how to maintain the world's fertility, including their own. With each generation, conception and childbirth become more complicated for the epic's royal characters, the seams more plentiful and vulnerable. Children are born in increasingly bizarre ways, and in each case, the manner of their birth mirrors a destructive delusion that will push the society closer to collapse.

I would like to retell some of these birth stories and describe the cycle that they communicate to me – the cycle which begins with the king's question to the river, and ends with the deaths of most of the civilisation's members.

After the river goddess leaves, the king marries a human being – a fisherwoman's daughter, another great beauty – to continue his line. A few generations follow, and eventually the family divides into halves led by the two brothers, Pandu and Dhritarasthra. It is with the Kauravas, born of Dhritarashtra, that we first see that something has changed with the human relationship to fertility.

Gandhari, Dhritarashtra's wife, will conceive and deliver her children differently than any other woman in the epic, and as a result have an entirely new kind of child. In her girlhood, Gandhari subjects herself to fierce penance – long periods of meditation and fasting – and in the old stories such acts, even when done in a power-hungry or malicious spirit, cannot be ignored. Gandhari, from before her marriage, has accumulated a reservoir of what could be called spirit-power. When she marries Dhritarashtra, who was born blind, she blindfolds herself so that she will not be able to enjoy anything that is denied to him.

Her power grows. Through her devotion to her husband as well as to various holy men, this energy eventually reaches a certain level, and Gandhari is granted a boon. She tells a holy man that she wishes to be the mother of a hundred sons, and her wish is granted.

It is an unprecedented request. One wonders if she is ever frightened — she has asked for more children than any human being has ever had, more than even the endlessly fertile Ganga had in eight years.

Gandhari's belly is full for nine months, then a year. She keeps running her hands over the skin, searching for the life that must be there, but there is nothing but a dull, immovable weight. In the meantime, she hears that her sister-in-law Kunti's first child, Yudhisthira, conceived with the god of death, is resplendent and beautiful.

A second year goes by and labour has still not begun. In jealousy and impatience, Gandhari hits her own womb. A spasm of pain grips her, there is a sudden tearing, and labour begins, but what comes out is not a child. It is described as a hard mass of flesh — featureless, lifeless, like iron. It lies on the floor, inert.

Horrified, Gandhari calls for the sage who promised her these sons, and tells him to look at what has emerged. Where are her children? The sage assures her that his promise will be kept, and begins to divide the lump. Does he use a tool, some kind of blade? It seems like he must, if the flesh is as hard as described. The lump must be cut or chiselled apart, forcibly torn. At the end of the process, the sage is left with a hundred pieces, each about the size of his thumb. He calls for a hundred pots. Each piece is placed inside a pot and then covered with *ghee*, the clarified butter still used for cooking and worship. The pots are placed in a carefully guarded spot and allowed to sit.

Today, I imagine these pots as they were drawn in one of my childhood comic books, all identical, lined up row upon row. The hard dots of flesh drink up the fat around them slowly, like chicks gestating inside their eggs. They will sit in the pots for two more years.

The epic is not explicit on this point, but over these years the pots must need to be cared for. The fat would be absorbed, the moisture would evaporate — more must need to be added, the flies driven off, and the level of ghee maintained. Gandhari cannot do this easily because of her vow of blindness. The growing embryos are too large a job for one person and must have become a group responsibility.

When more ghee is poured in, someone – a servant, a caretaker – would have looked inside the pots and seen, cloudily, the division of the cells, the transformation of the lumps from something featureless to salamandrine, a glance inside the mystery that had always been denied to humans about the nature of their fertility.

I can dimly see, buried inside the story, the birth of agriculture – the dropping of seeds into little holes, the tender nursing, the gathering of a crop – and also its inevitable development into something mechanical and extractive. In the story of the children and the pots, a new relationship to procreation is apparent – the grasping of the reins of fertility from the hands of the elementals, and its replacement by a more human-driven vision requiring more resources and manpower: a tribe of people manning the pots, dropping in the seeds, and eventually reaping an enormous harvest.

This is the result of division, and of the knowledge that is born when one asks questions – unprecedented bounty. Gandhari gets her hundred sons, and they are all healthy – an incredible feat in an era when parents must have been resigned to the deaths of a number of their infants.

Evil omens accompany the birth of the first child, who will be known as Duryodhana – 'he who is difficult to conquer'. Wise men tell Dhritarashtra to abandon the son, who they say will bring ruin on the family. The old, blind king cannot do it; throughout the epic, he is described as excessively attached to his son, but who can blame him? Is it more admirable to leave the child in the forest to die, as the sages recommend?

Another story, from earlier in the epic, hints at the source of these omens, and what might be wrong with children created through this form of fertility. Again, it features the abandonment of an infant.

In another kingdom, a childless king goes to the forest with his two queens. He begs a sage for a male child, and when a mango falls near the forest hut, the sage gives it to the king and tells him to feed it to his wife.

Having two wives, the king divides the fruit and feeds half to each wife. They soon become pregnant, but what they deliver is monstrosity instead of joy – halves of a child, kicking and screaming with hideous half mouths (the baby is split down the middle). The king orders the baby taken to the forest and left, as the sage advises Dhritarashtra to do with his eldest son, but the bundle of halves is discovered and, when picked up, fuses together. The joined child grows up to be a great king named Jarasandha, who can defeat even the tribe of the god Krishna in battle. It is only when Jarasandha is torn in half during a wrestling match, with the parts thrown in different directions, that he is finally defeated.

Until that moment, though, the divided man is stronger than those around him. Wholeness, when facing such an adversary, is a liability. In the *Mahabharata*, such people are defeated only by adopting their tactics – and in the process those same divisions spread within yourself and the society you will come to rule, leading to an end that is only postponed, not avoided.

Like Jarasandha, Duryodhana proves to be a man with a seam inside him, whose power comes partially from his brokenness. His overriding desire is to unify the kingdom and give the land a uniformity that he feels missing in himself. He can achieve this political substitute for wholeness only by taking away land from the rest of his family, and provoking the war that will destroy his clan and his society.

Brokenness is the central characteristic of many of this generation's rulers. Men with seams are able to divide life into portions and sequester certain parts – their faith, for example – into corners where they no longer make a difference. Jarasandha, for example, also born from pieces, is profoundly devout, but his devotion is limited to ritual and has no restraining impact on his behaviour, which involves the kidnapping of neighbouring kings and the absorption of their kingdoms into his own – again, a hopeless quest for a lost internal unity. Duryodhana, who continuously chastises others for failures of duty that he has himself committed (he cheats at dice to win the Pandavas' kingdom), cannot notice inconsistencies when they conflict with his desires.

If Duryodhana is a divided whole, his brothers prove to be something worse – little pieces of people. The 99 other sons who emerge from the pots are named occasionally in the epic, but not one has a personality or character apart from their oldest brother. With mechanical fertility, true individuality begins to leave the world. A group was required to bring these children into the world, and they appropriately end up feeling like constituent elements of a single creature, blindly following the commands of their older brother. There is a famous image from the stories of Krishna's youth of him dancing under the snake god, who forms an enormous hood, consisting of many snake heads that coalesce into a single enormous body; this is also the snake on which Vishnu reclines. The gods often have a single body and many faces. When humans try to achieve godlike power, though, they achieve it through many bodies and a single head.

Their power, though, is fragile, and the fragility can be seen in the nature of Gandhari's request. Gandhari does not ask for a hundred children, but a hundred sons, a request that an earlier generation would have seen as insane, because it throws the community out of balance. How can one find wives for this many men and ensure the maintenance of fertility, which requires – at the very least – two sides? (The epic is full of far too many male children, which is all anyone seems to pray for, and the five male Pandavas end up sharing a single wife.) No shepherd would ask for a hundred billy goats, or a hundred bulls. After they were slaughtered and sold, or worked the fields for a few years, there would be no offspring to maintain the increase, and the bounty would last for only a single generation.

This is exactly how long it does last: all hundred of Gandhari's sons are killed in the great war. Only one member of the generation survives – a half-brother named Yuvutsu. During Gandhari's long period of gestation, and the two years in the pots, Dhritarashtra impregnates his maid, who is delivered of a child in the ordinary way. As the generations progress, the aristocrats are increasingly incapable of procreating with each other. The children of the king who is abandoned by the river goddess – the sons of his human wife – are too sickly to produce offspring, so their mother must get a holy man to impregnate their two queens.

One of the ensuing sons is Pandu, who labours under a curse that he will die as soon as he has sex; his first three sons, the heroes of the epic – Yudhisthira, Arjuna, and Bhima – are actually born from his wife Kunti through direct congress with the gods of death, thunder, and the wind. When healthy children are born without the intervention of the gods, it is almost always through contact with someone from the margins of society or the lower strata – a fisherman's daughter, a wandering mendicant, a servant, someone in closer contact with the realities of work and the natural world. When all of his half-brothers are dead, Yuvutsu will rule what remains of the kingdom.

When I see the bent world's stretches of machine-flattened land, the vast rows of seedlings and circulating water wheels, the sows in their numbered farrowing crates, or the refrigerators full of our own chilled human eggs, ready to be implanted, I think of Gandhari's rows of pots, all bearing a single kind of child. It represents for me the birth of a mindset that – with the right methods and enough ingenuity – there is no amount that cannot be harvested.

There is another generation that follows, though, before the end comes for this culture, with deeper delusions. One knows that the end of a cycle of civilisation is approaching when the relationship with fertility takes a particular turn, and the epic tells us a story about what it will look like, marked again by the birth of a strange child.

The story concerns the end of the Yadavas, the tribe led by Krishna. When the two wings of family assemble for war, Krishna presides over the destruction – even though, as promised, he remains mostly outside of it, and only provides counsel and his services as a charioteer to Arjuna. The purpose of his avatar is explicitly stated as lessening the burden of the planet, which is dying under the weight of too many human beings. 'I cannot leave before their destruction is complete,' Krishna says of his own tribe in the *Uddhava Gita*, 'or they will surely overrun the earth.'

It is on Krishna, rather than any of the Pandavas, that Gandhari turns her rage at the end of the battle in which her hundred sons are killed. She curses him for allowing the slaughter to take place, and for repeated breaches of conduct in the war. She is entirely justified. Many of the worst instances of treachery in the battle are directly suggested by Krishna. At the outset of the conflict, the *Bhagavad Gita* relates his advice to Arjuna, the great Pandava warrior and son of the god of thunder, to fight despite the latter's desire to give up his claims to power rather than murder his own family, which would seem to be an entirely admirable impulse.

One can, in the end, get no firm principles of behaviour from the actions of this god, unless one keeps the whole in mind: the earth's need for fewer people will be satisfied not through moral exhortation, which is apparently pointless, but by encouraging people to destroy themselves. 'I am the knowledge of the Vedas,' Krishna will say, at the end of his life, 'and the refutation of that knowledge.'

Gandhari, standing in a field of her own dead children, cannot embrace such paradoxes. Wild with grief, she tells Krishna that, since he was indifferent to the slaughter of this clan, he will be slayer of his own people. Krishna smiles faintly. He says that he is himself bringing about the end of his people, and that her curse will only help him accomplish this task.

Thirty-six years pass. An exhausted peace has settled over the country. Yudhisthira and the Pandavas rule in the wake of the victory, and there is prosperity in the kingdom. Krishna's tribe, the Yadavas, fought on both sides of the war, and no more honourably than anyone else. The rules of battle were ignored day by day until none remained, and all of them participated in the closing atrocities, where men were killed unarmed, sleeping, and even praying. There is no atonement for such crimes, only distraction, and this is the course the tribe has taken.

In the years since the battle, the Yadavas have become vain and arrogant, addicted to drink and intoxicants, and fond of all varieties of luxuries. Their conquests over the land and resources of others have gotten quite easy, and with no true opponents left, they have oddly become both overambitious and lazy. If the tribe's eyes were still open, they might have seen a curse spreading through the world around them, the curse that was first pronounced by Gandhari: the great rivers, it is said, had begun to run backwards, and day after day the disc of the sun was covered with dust. The patriarch of these people, Krishna, has stepped aside, and is simply observing as Gandhari's words move through the water and sky.

One day, some holy men come to visit the Yadavas, whose leaders have more and more come to resemble uncorrected adolescents. As a joke, the tribe's young leaders dress up one of the men, Samva, in women's clothes, and paint him up, they think, very cleverly. Beneath the sari and jewels, they tuck some cloth to make him appear pregnant. Mincing their way to the river with great delight, they fall at the feet of the holy men and ask for a favour – can the sages predict the gender of the baby that this young woman will be having? She is the wife of a great warrior, they say, who wants very much to have a son.

The sages look down at this group of idiots, sprawled on the ground in gestures of comical devotion. These are the tribe's young, its future: the man in his ridiculous costume and his snickering friends.

The holy men know the name of the Yadava dressed like a woman – 'you, Samva,' one of them says, with growing rage, 'you will indeed bring forth a child, but it will be neither male nor female, but a rod of metal which will bring about the destruction of this race.'

The young men continue giggling, but nervously, while the holy men rise and leave. The sages go to tell Krishna of the curse they have pronounced, and he tells them, simply, that that which is destined will surely happen. The young men lie by the banks of water, resting. Slightly embarrassed, Samva begins to remove his clothes, and pulls aside the bundle that formed the false child. His hand grazes his belly and it feels suddenly taut. He presses with his hands on the spot and feels a pressure on his navel, then discomfort, and soon a tearing pain.

The men of the tribe gather around Samva as he writhes on the ground. Some of them laugh, wondering whether this is part of the joke. There is only horror, though, when Samva delivers himself of a large rod – slightly heavier on one end, like a mace or a club. Now it lies by the banks of the river. Like the children who are the partial offspring of the gods, divine curses appear instantaneously, with no period of gestation.

What does the rod look like? Is it smooth or rough, black like iron or shining like bronze? The word in Sanskrit, *ayas*, is the root of the word *Eisen* in German and iron in English, but it is also related to the word *aes*, bronze in Latin. I imagine the rod, though, as black and rough, a kind of pig iron, potential rather than a finished product – waiting to be refined, which will only happen over the course of the story.

The men carry their cursed object with them. They avoid Krishna, and instead take the rod to Ugrasena, their current king. Following the tradition of all panicked leaders, the king orders that the rod be ground into powder. From Dwaraka, their great fortress by the sea, the Yadava men cast this powder into the water. Sensing danger for the first time in many years, the tribe bans the brewing of intoxicating liquors and substances. They declare that people found making such substances will be impaled alive along with their families. It is a fitting decree: it is clear that they are no more noble in terror than they were in complacency.

The metal powder begins to sink in the waves and is soon caught by the currents. One larger piece is swallowed by a fish, and the other particles dance through the immensity of the oceans. Eventually the metal comes to rest on the muddy banks of the tribe's sacred waters. Perhaps it floated straight from where it was cast into the water, but I prefer to imagine the particles traveling around the world, passed from the hands of one sea to the other, and finally resting on the bank through the old law by which our poisons return to us.

On the earth where the metal powder lies, thick grasses spring up. They rustle dryly in the breeze, and there is a faint sound, difficult to place, as they sway, something like the sharpening of blades.

What in the world does this story mean? A man dressing up as a woman and becoming pregnant, and then delivering a metal rod: I remember being very confused by it when I was a child, particularly by the mechanics of how this rod came out. More importantly: why is this the means of destruction for the tribe? By putting this episode next to the other strange births in the *Mahabharata*, though, a kind of progression emerges.

The monstrous Jarasandha is born through a single mango, fallen from a tree. It is recognised as a gift, but is simply misused through an accident. Kunti's three sons are born through direct contact with the gods, and their two brothers are born when their father Pandu loses himself in the ecstasy of the spring and makes love to his second wife (he dies soon after, as the curse said he would).

Even Gandhari needs not only her husband and a crew of servants to create and care for her children, but also the *ghee* and the clay vessels that contain them. All of these parents can see that to have a child is not a human matter alone, but participates with the earth and comes finally from an ultra-human source.

My wife and I were looking at our son the other day, and a thought crossed our minds, *I can't believe we made this.* But the true wonder is how small a part we play – his body is nourished by the sun and the food my wife eats, with the milk partially digested in his stomach by the bacteria that colonised him the moment he emerged from the womb, and his life and development are guided by a set of instructions (how inadequate that word seems!) that have been enfolded in him over the course of generations, human and amphibian and bacterial. Every living thing, and every unliving thing that rests until it is gathered into the stream of life again, comes finally from some shining, unfathomable source.

To dress up a man as a woman and pretend that he is pregnant, though, is the ultimate image of a sense of *self*-propagation. It is the final stage of the journey before collapse – the delusion that one can produce wealth, offspring, fertility, from nothing but oneself. A community that is tipped entirely toward one sort of person – men, for example, or warriors or engineers – will start to believe that no other kinds of people are necessary. When great wealth is created through technical means, as Gandhari's children are created through the pots, the resulting generation becomes convinced that technical means are all that is required: perhaps sons can come out of other sons (remember that the Yadava men tell the sages that they are hoping for yet another boy). Self-propagation is a kind of outer limit of delusion, and it is neither so far-fetched nor far off in our society: we are already starting to explore whether meat might be cultured out of inanimate materials and thin air, and plants grown entirely out of synthesised vitamins and ingenuity. I await the day when, through the blind advance of DNA research, someone tries to have a child purely out of their own genetic material, unwilling to get anyone else involved. I suspect that the meat and plants will be curiously tasteless, and the child very strange – all harbingers of the end which comes for the Yadavas after the birth of their iron son.

As the grass grows by the bank, filling up with sunlight and drinking the metal at its roots, strange omens appear. In clean food that has been carefully cooked and prepared, live worms are found. Birds call constantly, and not just at their appointed times in day and night. While men in the city sleep, mice and rats come and eat away their nails and hair. Everywhere, life is in great confusion, especially in the animals domesticated by humans: asses are born of cattle, elephants of mules, and cats of dogs. One senses that, since the question was first asked of the river goddess, men have gradually lost all sense of how to channel fertility.

Rather than making the Yadavas fearful, these omens remove their last restraints. They cease making offerings to the gods and holy men, and hus-bands and wives indiscriminately deceive each other − all ideas of fidelity begin to seem ridiculous. Krishna tells the tribe that they must make a pilgrimage to their sacred waters. On the banks near the grass, they form their camp. An immense feast has been brought, and with the mysterious agency found in old stories, alcohol has found its way into the carts and bags. No-one is blamed for its appearance; its presence is not even noted.

The Yadavas begin to drink. Talk turns to the great war, and they begin to accuse each other of crimes − killing soldiers while asleep, even beheading a praying man. Soon, the argument turns violent and one Yadava warrior is killed. A melee follows. 'Impelled by the perverseness of the hour that had come upon them,' the Ganguli translation reads, 'all became as one man' − a single head with many bodies. Krishna sits calmly and watches them. He reaches over to the grass growing by the bank, lifts it up, and scatters it at the feet of his tribe. The fighting men take up the grass, hitting each other with stiff pieces, and discover that each blade of grass has taken on the properties of the metal bolt. Men struck with it fall dead; the rushes can penetrate armour, lodge in bodies, lop off arms and legs. Gentle nature has lost its pliability, its willingness to bow to the demands of men.

Krishna takes up some of the grass, and as his kinsmen fight around him, he begins to kill them. No-one tries to run away − they keep rushing at each other as the warriors did in the great battle. 'As rivers flow into the ocean, all of the warriors of the world are passing into your fiery jaws,' Arjuna says in the *Bhagavad Gita*, 'all creatures rush to their destruction like moths into a flame.'

Finally, all but a few of the male Yadavas are dead. Krishna sends two messengers to tell Arjuna, the great warrior whose chariot he drove in the battle, what has happened, and to come protect the women of the town. Then he walks into the woods. It is here, in some versions of the story, that he returns to his four-armed self. In two arms he holds wrought objects, both man-made weapons, a golden ridged disc and a simple metal mace; and in the other two arms is the world which civilisations neglect as they become enchanted with artifice. On that side, Krishna holds the conch shell in one hand and in the other, *padma*, the lotus, ancient symbol of balance. On one side, objects extracted from the earth and shaped by men; on the other, gifts – both edible and beautiful – from the world of the water.

Sitting against a tree trunk deep in the woods, Krishna moves his foot slowly back and forth, and a hunter named Jara mistakes it for the ear of a doe. He takes out his bow and shoots an arrow. At its tip is a piece of metal the hunter found in a fish, the last and largest piece from the metal rod. As Jara follows his arrow, he is horrified to find Krishna sitting calmly, the blood pouring from his foot. The blood is said to be astonishingly beautiful on his skin, like the blossoms of flowers. The hunter falls at the feet of the wounded god and begs for forgiveness. 'Get up and do not fear, O Jara,' Krishna tells him, speaking the last words of his avatar. 'This deed was actually desired by me.'

With Krishna's departure, the Kali Yuga, the dark age, begins. The cycle, in one sense, is complete. But the epic also shows the turning of the wheel, how it returns to where it began. When Arjuna travels to the city of Dwaraka, the Yadava capital, he finds it full of weeping. He arranges for cremation rites for the dead. Many of the widows throw themselves onto the flames.

Weapons in hand, Arjuna leads the enormous procession out of the city. The people – wives, children, the old – have gathered what wealth they can, but much of it has to be abandoned. As they walk out, the waters begin to rise. The flood passes through deserted streets and into homes, sweeping the work of their lives into the current. Beholding the sight, the remaining Yadavas speed up their walk, saying 'wonderful is the course of fate.'

From the city, they enter the forests, looking to found a new capital. They have entered a lovely rich land, full of animals and food. This train of old men and women, travelling through wild country, is fearless because they are protected by Arjuna, who has never been defeated in battle and possesses divine weapons: a mighty bow named Gandiva and an inexhaustible quiver of arrows.

Arjuna's reputation has not spread to the forest, though, which is already full of people who live off its bounty – frugally, clearly, because the bounty remains to impress visitors. The forest people also rob unsuspecting travellers, and cannot resist attacking the train of women and goods protected by a single man. The text calls them *Mlecchas* – non-Vedic people, or (in the closest English equivalent) barbarians.

The Mlecchas scream and rush out at the army with clubs. Arjuna warns them to retreat, and when they do not, takes out his mighty bow, which for the first time in his life feels heavy in his arms. He finds he can barely string it. When he begins to shoot, it is only with difficulty. One forest man falls, then another. The arrows speed from their bow, still finding their mark. Then Arjuna reaches down and finds that his quiver, for the first time, is empty. He tries desperately to beat the men back with his bow. It is useless. Finally, he stops altogether and watches the last wealth of Krishna's tribe being dragged away.

At first, the Mlecchas drag the Yadava women into the woods. When it becomes clear that Arjuna is powerless, these women, husbandless and without protection, begin walking of their own free will into the woods with the attackers.

It is an extraordinary moment in the world's ancient literature: the willingness to acknowledge the senescence of a great warrior and the civilisation to which he belongs; to describe the glory of its cities but know that this glory cannot be maintained, because it will eventually decay the land and all its creatures, and to see that fertility must move back to the forest where the goddess once lived. It is the same largeness of mind that Simone Weil found in the people who produced the Iliad – the ability to see war 'unveiled either by the intoxication of pride or of humiliation.' One suspects that, as the *Mahabharata* was gathering together in the society's imagination, the songs of the vanquished mixed with those of the conquerors, producing one of those rare stories which, if somehow held together in the mind, become the whole of existence: rise and fall, birth and death, endless ambition and the absence of all desire.

The forest people are not rapacious; they do not kill every Yadava or take all of their wealth, because there is enough left for the survivors to found a new city on the banks of the Sarasvati River. It is a small city, less luxurious, easier to maintain. Many of the survivors, nonetheless, have no interest in staying there. The epic says that they go to the forest, eat roots and berries, and live out their days in contemplation of the god who has just departed.

Arjuna is not ready for such devotion. He has been defeated, and not by another great warrior but by a horde of club-wielding forest people. In distress, he travels to the home of the sage Vyasa and tells the holy man the story of his humiliation. The sage offers no pity. He tells Arjuna that his weapons, along with his understanding and his prowess, have never belonged to him, and he has no particular right to be upset. Dhananjaya is one of the names for Arjuna (it means 'one who conquers riches'). This is what Vyasa says:

All of this has Time for its root. Time is, indeed, the seed of the universe, O Dhananjaya. It is Time, again, that withdraws everything at its pleasure. One becomes mighty, and, again, losing that might, becomes weak. One becomes a master and rules others, and again, losing that position, becomes a servant for obeying the behests of others. Thy weapons, having achieved success, have gone away to the place they came from. They will, again, come into thy hands when the Time for coming approaches.

One is reminded of the Yadavas as they contemplate the immersion of their city – 'Wonderful is the course of fate.' What passivity, one might say, what determinism – or, to use a term which one Western observer from a more confident era called 'the running sore of all Oriental countries,' what *quietism*. One expects wailing, grief, or at least the learning of some hard lessons. Perhaps this enormous ability to accept encouraged them to do nothing while they were creating a catastrophe. This is the sensible view.

The *Mahabharata* tells a different story, and increasingly it feels more true to me, and certainly more useful in times of disaster. The Yadavas waste no time in breast-beating or recrimination, although all of them ignored warnings and participated in numerous crimes. They are able to move on to the next stage of their lives without spending energy on regret. Their descendants will probably make certain mistakes again, in slightly different clothes – that's just the way it is.

Perhaps there is no way out for the society, but there is a way out for individuals. At the outset of the epic, two aged queens – the daughters-in-law of the king who asked the river goddess his question – are visited by Vyasa. 'The past has gone by pleasantly,' Vyasa says, in Rajagopalachari's translation, 'but the future has many sorrows in store. The world has passed by its youth like a happy dream and it is now entering on disillusionment, sin, sorrow, and suffering. Time is inexorable. You need not wait to see the miseries and misfortunes which will befall this race.' The queens enter a hermitage in the forest and die rapt in meditation.

This is the escape offered in the Hindu tradition, and I suppose several others as well. It is always possible to exit the cycle; there is a forest some-where, although they are getting harder to find today, and will not be as rich in life as they were once were. Still, they exist, or will come back; the society will collapse before it can destroy them completely. If anyone wishes, they can enter the stage of the cycle that will be eventually be forced on them, but without the accompanying sorrow; they can cultivate the understanding that exists in the waters on either side of the epic. Perhaps it feels like a terrifying blankness, an effacement of self, but in it are the seeds of rebirth.

All of the war's survivors eventually end up in the woods. Gandhari, Dhritarashtra, and Kunti will go to live there, and when they see a fire approaching, sit down in meditation. They are swallowed up by the flames. The Pandavas, when they hear of the destruction of the Yadavas, abandon their kingdom. As they leave, they give away their jewels and rich garments and wear the bark of trees, walking through the world with their wife Draupadi and a faithful dog. It is on this journey that Arjuna will finally give up his weapons, throwing them into the sea. Standing on the cliff above the water, they realise that this is where Dwaraka, the great fortress of the Yadavas, once stood. Now it is entirely covered by a 'sea of red water' – the blood of the dead, perhaps.

Soon after the weapons are taken by the waves, the party will continue walking, and one after another they will die. The civilisation they leave behind will be much smaller, and beginning cautiously again in the forest. Nothing is entirely destroyed, there is no total apocalypse, and it is clear the cycle is beginning again, although the cycles of the Kali Yuga will have their own shape and speed. Nonetheless, the outlines seem to be the same. Human beings receive gifts – weapons, children, prosperity – and in the absence of the departed gods begin to feel wrongly that they are in control of them, and finally that they, the humans, are themselves the source. The immense harvest that they have extracted from the world through their sophistication – the wrought metal, the shining palaces – begins to disappear as the authentic sources of vitality are lost or forgotten. Then the flood comes – or the ice, or the fire, or the men with clubs; they destroy the harvest and the palaces, but somewhere in the ruins the true spring is found, and men can kneel down again by the water and drink from it.

Two last stories, from the halves of my family (one needs both, or there will be no healthy offspring).

First, from my father. When the Americans were about to land on the moon for the first time, he was a teenager in India. Before the space shuttle landed, he asked his grandmother, teasingly, the question of a divided man, about whether the American astronauts would find our god there (the moon is a minor deity in the Hindu pantheon). He has never forgotten her reply.

No, his grandmother said, they would find nothing there – the gods leave when they find out the people are coming.

The second is from my mother's side of the family. A year after her father, my grandfather, died, we went to Varanasi to perform a ceremony for him. I know no Sanskrit, and felt very little during the days of chanting and rituals. Even the priests seemed bored to me. We ended the trip with a bath in the Ganga. The guidebook warned us not to do it – too much sewage flows into it now, the river is filled with coliform bacteria. We did it anyway. As I was looking at the flow of the great river, ready to submerge, I felt something immense moving along with it, something larger than the water or the people washing their clothes on the banks. It was unmistakably there. I threw myself into the muddy, contaminated water to escape from it, and when I came back up it was gone.

I have felt something like this a few times in my life and each time I have cast it off. I dismissed the feeling later, wondered if I was somehow putting myself on. I wonder that even now. I read an aphorism later by the poet James Richardson – *you will know the true god by your fear of worshipping it*. I am not ready for this gift yet – clearly I am not, because I am writing about it, asking questions, providing explanations. The gods are right to leave in the face of certain kinds of searching. Eventually, though, when your society has been humbled into prostration – and me too, I am not nearly there yet – the deity returns, the river begins to walk through the world again.

References

Except where noted, the quotes from the *Mahabharata* are from Kisari Mohan Ganguli's 19th-century translation of the epic. It's available for free online. The place for curious English speakers to begin is probably Rajagopalachari's abridged translation.

The *Uddhava Gita*, like the *Bhagavad Gita*, is a later spiritual text grafted onto the narrative of the *Mahabharata*; it consists of Krishna's last remarks to Uddhava, a surviving Yadava, before Krishna departs the earth. In the essay, the quotes from the *Bhagavad Gita* are from Eknath Easwaran's translation (Nilgiri Press, 2007); those from the *Uddhava Gita* are from Swami Ambikananda Saraswati's translation (Ulysses Press, 2002).

The line from Simone Weil is from 'The Iliad: Poem of Might,' which I found in *The Simone Weil Reader*, edited by George A. Panichas (McCay, 1977). The Western observer from a 'more confident era' is Count Gobineau; I found the quote in the introduction to Gertrude Bell's translations of Hafez. Finally, the aphorism from James Richardson is from his *Vectors* collection, which I found in *Interglacial* (Ausable, 2004) – when I looked it up, I found that I had misremembered it slightly, but felt closer to my incorrect version, so I let it stand. The original is, 'you will know the real god by your fear of loving it.'

Openings

Freeing Space for a New Cosmology

TIM FOX

Issue 5, Spring 2015

I remember the rocket.

The house where I lived in 1972 was about nine line-of-sight miles from Cape Canaveral. On December 7th of that year, in the heart of the night, my family gathered with friends and neighbours to watch the massive Saturn V rocket carry Apollo 17 – the final moon mission – into history. We huddled around the TV up until the last minute of the countdown, then my dad hoisted his two-and-a-half-year-old son onto his shoulders and led the way out onto the lawn. There we stood in awe and silence as the fiery dart rumbled into the stars. Even from our distant vantage, I could feel the earth quiver.

Forty years later, the old memory rises to the surface as I pull another thin paperback off the shelf in the little two-room O'Brien Library, tucked among towering Douglas firs, incense cedars and big-leaf maples on the outskirts of Blue River, a struggling rural community in the central Oregon Cascades. The library is stocked by donation, staffed by volunteers and has been running check-out and returns on the honour system for over eight decades. Today is my first shift in the stacks, and my first job is to cull the science fiction section and open up space for new additions to the collection.

On the front of the paperback I hold is a shiny silver rocket, rumbling away from Earth toward a tiny red dot. A thin grey dust layer fuzzes the top of the book. When I lift the cover, it cracks loose as the brittle desiccated spinal glue gives out. Inside, the yellow pages are splotched with black stains that look like a defunct bacterial colony left too long in a forgotten Petri dish. The fine print says this particular novel – a saga of human glory set in a Martian outpost – rolled off the pulp press in the late 1950s. By all appearances, it has not been opened since the Kennedy administration. I lay the front page flat and slam down the stamp of judgment. 'Withdrawn'. And into the discard box it goes.

By the time my three-hour shift ends, I've worked my way up through the Ds and passed sentence on dozens of titles, an inordinate proportion of which fall into one of two categories: the hypertechnic exploits of an inter-galactic humanity, and post-apocalyptic home world nightmares. Taken together, the effect seems almost conspiratorial. How else to explain the widespread bias for painting dead and distant worlds in the rainbow colours of promise while this wild, verdant, beautiful Earth, the only gem of life in the known cosmos, is rendered with a palette of shadows?

Perhaps, rather than conspiracy, this small sampling reveals a common cosmology that has, for decades, captivated not only society at large, but a whole legion of science fiction writers who have expressed it through their efforts to rouse civilised excitement about the colonisation of other planets. In particular, Mars, the most promising of the lot (owing to its relatively close proximity and the presence of life's most essential ingredient, water). These writers, and their dystopian doppelgangers, along with scientists and politicians and people from all walks in between, have been so successful that the fourth planet has been deemed worthy of spending billions of dol-lars on real probes, landers and other visitation devices, each an automated vanguard in an effort to one day stamp human boot-prints in the red dust plain even while our own world turns to dust.

It's quite an achievement, spinning a frozen, barren, desolate, inhospi-table planet millions of miles away into a possible home, while concurrently spinning our beautiful living home into a hell. And the net effect is blind-ness to the fact that the earth under our feet, even now, in the throes of so many socio-ecological crises, has much more going for it than Mars ever will. Antarctica looks positively tropical by comparison.

What is puzzling is how the collective consciousness came to this blind-ness in the first place.

I think it has to do with a particular criterion the cosmology of domi-nant culture has long used in determining what constitutes the so-called good life. That largely unspoken criterion is not the successful long-term inhabitation of a homeland, but expansion beyond it; growth, progress, advancement. And Earth, though still humbly habitable, is a world on which the possibility of expansion is nearly played out.

But Mars – a whole planet virtually untouched, and theoretically within reach – offers a chance, slim though it is, for us to remain in the habit of expansion. We have only to sell out this planet in an all-or-nothing gamble for the war god's favour. That, after all, is what expansion has always been about, ever since our direct cultural ancestors – the agriculturalists of the Fertile Crescent – exhausted their home soil some six millennia ago and faced a future of limits and reduced numbers or a future of conquest and continued excess, which was, even then, the definition of prosperity among the elite minority who benefited most from it. And so, conquest it was. War.

Such aggression required extreme rationalisations. That task fell to the official storytellers of the day (the priesthood), who began spinning the glory of Mars even before the Romans had given the god his name. Even before Rome existed at all. The force – the spirit – that Mars came to mytho-logically embody was the colonial conquest and control of others. His spirit was, some six millennia ago, a new force loosed upon the Earth and with it came an unprecedented shift in cultural consciousness, away from an emphasis on ever-deepening integration into local landscapes, toward an emphasis on the golden riches to be found on the other side of an apparently inexhaustible frontier.

That frontier has been spreading outwards from the perceived insular centres of civilisation into a perceived ocean of untamed wilderness ever since. The whole time, the frontier has been a margin of conflict, of colonial aggressors waging a genocidal, ecocidal, even geocidal campaign of theft, subjugation and replacement against the human and non-human lives already dependent upon what the invaders see as their rightful spoils. On a deeper level, the invader incursions represent the replacement of countless grounded cosmologies with an increasingly singular cosmology rooted in longing. And the farthest conceivable reaches of the invader's longing is the sky.

It's no wonder, then, that our cultural ancestors eventually imagined the sky as the dwelling place of their gods. For them, divinity found its source in the dependability of the heavens, up and away from the increasing mess-iness and unreliability of earth, never mind that the biotic abundance on which human corporeality and wellbeing depended originated in the soil's translation of sunlight into life.

The invader's mode of existence mined the soil. So the soil continually failed them. It couldn't be trusted. And everywhere they went it reinforced its untrustworthiness by suffering eventual exhaustion and consequent poor yields. How, under these circumstances, could so feeble and treacherous a goddess as Gaia be revered? What the soil miners failed to see was that her apparent feebleness and betrayal did not derive from the earth, but from their own excesses. It derived from their increasingly alien relationship with the landscape. Rather than heal the relationship by relearning how to live within the limits of the earth – within the planet's annual solar budget – they turned their gaze upward and sought to emulate the gods above, gods of their own invention perched upon untarnishable golden thrones.

If the people could get better at bringing the lasting power of the heavens down, maybe the messiness and limits could be overcome. That became the programme the storytellers started to sell. The holy goal of godliness. In other words, complete control.

This went on for millennia and all the while the programme of control grew more refined and entrenched. The original kindred enspiritedness recognised as inherent in every star, leaf and breeze gave way to a multitude of anthropoid divinities – many still peripherally bound to the earth through the forces of nature they personified. Eventually, this multitude was further reduced to the equivalent of a spiritual wheat field, the monocrop of a single deity, singularly male, separate and above the mortal realm. He was a book-bound god, purely abstracted, and thus conceptually omnipotent. But what his believers overlooked was the real earthbound precondition on which this god's omnipotence depended. The frontier.

Wilderness, the ultimate foil for civilisation, had to remain the oceanic realm into which the conquering heroes and their armies could forever advance from their islands of civilisation. And that is how it was until 1893, when civilisation crossed a threshold and became the rising ocean surrounding now-shrinking islands of wilderness.

Historian Frederick Jackson Turner recognised this moment and expressed it that year by declaring the frontier closed. Actually, at the global scale, the frontier did not close in 1893 , but rather, it *began closing* for the first time since it opened some six millennia earlier. You could think of this event along the lines of the old riddle (slightly modified): 'How long can you run into the wilderness?' Answer: 'Halfway, then you're running out.' The year 1893 represents the end of civilisation's long run *into* the wilderness. After that, it began running out. At an accelerating rate.

But the implications of this new situation went unrecognised and thus, civilisation failed to undergo the essential corresponding social inversion (cultural deceleration, contraction and diversification: in other words, maturation). Instead, it continued to advance the divine programme of monocultural expansion and complete control, thereby becoming increasingly out of sync with reality. And now, barely a century later, the last wild islands are almost flooded. Expansion has again run up against the wall.

And eyes long turned skyward for divine inspiration now look that way with a different intent. Divine ascendance. That is the ultimate objective to which we apply our scientific curiosity and exploration. The methods of science (some would say a god in and of itself) remove moral considerations altogether and render the interplanetary colonial effort little more than a technical challenge. And so the scientific priesthood uncritically constructs and rockets mechanical missionaries into space to prepare the way for the next logical wave. Flesh and blood aliens.

Looking back, we can now see that the moon was a practice run. The colonisation of Mars represents the real deal, the culmination of the programme of complete control begun all those millennia ago. In fact, Mars is a world that will not accept us *unless* we are in complete control. Only as gods will we be able to exist there. Earth, on the other hand, can and does expose our hubris by resisting complete control in direct proportion to our every effort to take it. So we dream of a red heaven.

Well, not all of us. I, for one, am opening to a different possibility: withdrawal. Into places where long-latent spirits stir.

I slide the box of discards over behind the circulation desk for the next volunteer who will leaf through the card catalogue and pull the titles. As I'm putting on my coat to leave, I glance back at the shelves. Where I've been working, there are large gaps between the remaining volumes. The sight is somehow freeing. In the gaps, I see opportunity; what as-yet-unwritten tales might fill them in? I can't imagine, but I feel heartened nonetheless as I head for the door.

When I step outside into the cool fresh air, the palette of autumn draws my eyes upward not into the blue sky, but into a vision of vibrant yellow maple leaves in their full glory. They shimmer and seem to glow with their own inner light, offering a rich contrast to the deep greens of the stoic conifers who are cast in sharp relief by long October shadows.

Standing in the forest, I'm suddenly struck by the sense that I've just entered another library, a library of trees. The stories to be read on each leafy page would more than fill the openings. I imagine many of those stories would be about a long overdue homecoming set on a world that grows ever more wild, verdant and beautiful every day. A world with people struggling, longing, ceaselessly living to be grounded, integral parts of it all.

A breeze whispers through the canopy. Dozens of leaves release, each a golden spirit. They flutter down to earth.

As they fall, another memory rises, my first memory, deeper than the rocket.

In the heart of a cool summer night, in a green canvas tent set up in a Maryland forest on the other side of the continent, my mother is tucking me into a sleeping bag on the ground. I lay my head down. She bends, gives her two-year-old son a kiss then rises and silences the hissing lantern.

Beneath me, I feel the earth quiver.

FOUR

We will reassert the role of storytelling as more than mere entertainment. It is through stories that we weave reality.

KATE WALTERS — Feeding from the Fire Below — *Watercolour and conte on gesso-prepared canvas — Issue 6, Autumn 2014*

A dream of The Mother: a sense of being supported by the fruitfulness of an intense, trusting and yielding relationship with Nature, here embodied by a deer and an acacia tree; and the unceasing and yet-to-be-born which gives Mothering life. The support is implied by the figure's connection with a spirit animal — in this case the Horse, which according to Marie Louise von Franz represents the authentic 'voice of the cells' — in touch with the Fire which lights all life and is found in the earth.

The Unconscious and the Dead

The best classes always have somebody dying in them.

JOHN REMBER

Issue 4, Summer 2013

I

I had titled the course *The Unconscious as Literature*.

No matter that I didn't know what I was talking about. Any time you talk about the unconscious you can't know what you're talking about. That's why it's called the unconscious.

'What I want from you by the end of the semester,' I told the students assembled for the first class, 'is an understanding that it's not what you know, but what you don't know that is going to determine your life.

'At this point in your lives, you probably think you're conscious beings. You're dead wrong. Consciousness is only attainable after decades of being honest with yourself followed by more decades of honest observation of the world. Even then, consciousness is mostly illusion.

'Your consciousness isn't going to pick the person you marry. It's not going to choose what kind of sex you have and how often. It won't decide if you have children, or if you abuse drugs and alcohol. It won't determine how you talk to your parents, or if you'll graduate. It's your *unconscious* that will preside over these things, as well as what music you dance to, what corporation you work for if you're lucky enough to get a job, and whether or not you plagiarise the paper you write for this class.'

Not good news for my students, at least if they had believed me. Most of them were certain that by going to college they had achieved free will – and its necessary component, consciousness – for the first time in their lives. But I had been teaching undergraduates for fifteen years, and had plenty of evidence that none of my students –or colleagues, for that matter – had made the big decisions of their life from a place of much awareness.

I had taught people who had married abusers after childhoods with abusive parents. Some of my former students had realised they were homosexual after being heterosexually married for ten years. One had gone into the hospital for an appendectomy and come out with a new baby.

I had taught people who worshipped money when they thought they were being devout Christians. Others had gone into teaching because they had high needs for control, into politics because they deep down wanted to hurt people, or into medicine because they loved sickness rather than health.

I also had a photocopied stack of plagiarised papers in my office. I had handed back the originals to their purported authors, saying, 'You're way smarter than I thought you were. You get an A. That's got to feel good.'

That sort of thing looks cruel to me now, but at the time I thought it was a clever way of shocking people into an awareness of their own destructive impulses while I gave vent to a few of mine.

I need not have bothered. Over my career I watched as awareness snapped into being, even if was too late to do any good. I knew people who had died in Afghanistan or Iraq because of unspoken but real family expectations, or who had overdosed on drugs that allowed a glimpse of worlds they yearned for but couldn't live in, or who had killed themselves for love or the lack of it. People who had gone deeply into debt for medical degrees quit medicine and drove trucks or opened bakeries. People struggling through grad school became parents when they didn't want children and couldn't afford them, or got depressed or sick when the choices they hadn't made started requiring more of their lives than the choices they had made. Highly intelligent and literate colleagues smoked and drank and ate themselves to death.

Witnessing these things brought me to the hypothesis that if you don't continually work on becoming ever more aware, parts of the person you think of as you – maybe all of the you that you think of as you – will sink into a dream not too far up the continuum from death. A disturbing corollary to this idea is that if you try to go in the opposite direction, your unconscious will be exposed to the light, and it doesn't like the light.

'We will start with a few scenarios,' I told the class. 'One presumes the unconscious is just blind impulse from the parts of your brain inherited from fishes and reptiles, something to be controlled, if possible, by reason, so you can live as free and thinking beings, instead of just eating and shitting, spawning and dying. Another scenario is that the unconscious is an alien entity in your skull, one that wants to keep on living its unexamined life, even at the expense of destroying your awareness. Yet another is that the unconscious is the underworld, alive and stirring beneath your feet, full of paths not taken, unborn selves, unexpressed yearnings, and forbidden thoughts.'

II

A student's hand went up. It belonged to Jamie, who had been in several of my classes already. I'd given him an A in each of those classes, not because he was a plagiarist, but because every time I read one of his papers I realised he was smarter than I'd thought he was, and I'd thought he was smarter than me to begin with. He didn't have to be told that free will was mostly illusion, for one thing.

Jamie had been diagnosed with soft-tissue sarcoma when he was seventeen. He was twenty-two, having lived four years longer than he had been told he would live, mostly because every time a tumor showed up in his CAT scans, his doctors cut it out.

He was a shadow of his former self. He was missing lymph glands, big chunks of muscle, his hair, and portions of both lungs – those were just the parts I knew about. His ambition was to live long enough to graduate, but over the semester break he had sat down at a big table with his doctors, nurses, his physical therapist, the hospital chaplain, and a hospice worker. They told him that he wasn't going to make it.

'I have a scenario for you,' Jamie said to all of us. 'I've got three months to live. It's a lousy way to get out of a final. I'd appreciate it if you'd all make this class the best class I've ever had.'

We considered his words in silence. The classroom looked the same, but the walls, the furniture, and the language cast shadows they hadn't cast before. Even Jamie, as a shadow, cast a shadow. He still had to take a final, for one thing. Just not my final.

I spent the rest of the class going over the reading list. I had doubts that the class would be the best class of Jamie's life, but I thanked him anyway, for giving us a head start on dealing with the unconscious.

III

Jamie, dead for over a decade, has lately been on my mind. He doesn't speak to me in dreams, but he's started to talk to me when I sit out on my deck with a cup of coffee.

For a long time I didn't think I could talk to him because he's dead. That has proven less of a barrier than I expected these warm and smoky mornings of this year's fire season. Jamie and I have a dialogue of sorts, although he won't answer any questions about where he is now. Also, he issues disclaimers: 'If there's any of me left in your world,' he says, 'if I'm not just a hallucinatory caffeine overdose, I doubt what I have to say will be of interest to you. You're alive. I'm dead. Death is a total rearrangement of priorities.'

'That's why it was so good to have you in class,' I say. 'We couldn't watch you taking notes for an exam you'd never take and not be a little more thoughtful about what we were talking about. We thought you were getting something we were missing.'

'The best classes always have somebody dying in them,' he says.

'As long as it's not the professor,' I say.

He grins, and I suddenly realise he has no lips. 'Your turn's coming up.'

'After your chair was empty,' I say, 'we had a game: What Would Jamie Say? It fit in well with the other game we played in *The Unconscious as Literature*: "I Know What You're Really Saying and Why You're Really Saying It."'

IV

One of the books I assigned was Vladimir Nabokov's *Lolita*. It illustrates how human confuse unconscious imperatives with conscious decisions. Fifty-five-year-old Humbert Humbert, the novel's narrator, mistakes twelve-year-old Lolita for a mythical being wonderfully inserted into his gray existence, when in reality she's a little girl whose nascent self cannot bear his gaze, much less his touch. He kidnaps her anyway and takes her on a motel-to-motel road trip across America, where his possessive lust and tacky fear of growing old drags her from semi-divine nymphhood into an ugly and claustrophobic middle age.

By the end of the novel, Humbert Humbert is in gaol, dying, but Lolita has it worse. Once she existed in unthinking beauty and timeless grace, but she's become an ordinary housewife – Nabokov's word for her is *slattern* – forever exiled from the soft-lit green and gold and endlessly renewed world that nymphs inhabit, forever unable to connect with the eternal spark that once animated her. It's hard not to see Lolita's transformation as a death.

Humbert Humbert can't live in Lolita's pristine and youthful world, which is what he wants to do. Instead he pulls her into his own foul and decaying life. The lesson is clear: touch a goddess, and she'll turn into rotten flesh in your arms.

The old idea that humans die when they see or touch a god is exactly wrong. No god can experience the human touch and live.

138

'But there's another way to look at this,' I told the class, which at that point still included Jamie. 'Humbert Humbert sees Lolita as a metaphor that counters his fear of aging. But she's a literal human being as subject to decay and death as he is.'

Jamie spoke up: 'I'm not a person. I'm a metaphor.'

Other students immediately protested that of course Jamie was a person. He shrugged and didn't explain further until I got his paper for the class. He sent it to me from his hospital bed – it was seventy-one pages, longer than the twenty pages I'd specified. He apologised for not finishing it.

He explained his metaphorical status by describing his return from anesthesia after yet another operation:

> Slowly I piece together what has been done to my body. An IV in my right arm, an IV in my neck, an arterial in my left arm, a catheter in my penis, an epidural catheter in my spine, two chest tubes (one anterior, one posterior) going in between my ribs, an automatic blood pressure cuff that turns on by itself every ten minutes, six EKG probes on my chest hooked to a monitor displaying my vitals, a pulse oximeter measuring blood oxygen saturation on my right pinkie, an oxygen line in my nose, and two nylon bags around my calves that inflate and deflate to keep fluid from pooling in my lower extremities.
>
> The really bad post-surgical pain is four hours away. While I'm still floating I'll tell you about this place—the ICU, and Lucretia, my ICU nurse.
>
> Lucretia is a controlling, high-strung, by-the-book bitch who has never gotten over the fact that she's not a doctor.
>
> Lucretia knows me as '21-year-old male undifferentiated sarcoma, post-operative right lower lobectomy for chest metastases.' But she calls me Thoracotomy Bed Two, for short.
>
> Lucretia doesn't like me very much. It's not that Thoracotomy Bed Two is a bad person. But Lucretia has never taken care of a thoracotomy that wants to be a person before. Lucretia is [usually] in charge of bypassers and renal failures, who apparently don't care as much about personhood.
>
> Lucretia doesn't like having Thoracotomy Bed Two for another reason –there's little chance of my dying during her shift. She wants to have Angioplasty Bed Six, so she can get the

crash cart out when he codes in five hours. Lucretia needs a fix, and unless my heart stops, she's not going to get it.

I hate Lucretia. But when the pain comes I'm nice to her. She has the key to the narcotics drawer.

That's what Jamie wrote in the last semester of his life.

Poor Lucretia. Poor any nurse or doctor who has a patient who under-stands the violence that's been done to him and who has realised that much of that violence is not conscious.

The rest of Jamie's essay covers a wide range of experience, but if I had to sum it up, I'd say that it's the account of an aware human being trying to stay that way in the face of malignancies – his cancer, and the medical system it's condemned him to, and that system's less than conscious personnel.

V

Does our civilisation have an unconscious? That was the question I asked my students halfway through *The Unconscious as Literature*.

Every one of them asserted that our civilisation did, indeed, have an unconscious. They went further. Their families had an unconscious. The college had an unconscious. The local police department had an uncon-scious. The stock market had an unconscious. In the deep red state of Idaho, the Republican Party had an unconscious that was whispering to them about absolute power over everybody, like a shadowy, gleeful Mephistopheles whispering to Faust.

But by that time my students were seeing the unconscious in the toasted tortillas on their plates in the dining hall.

'If the unconscious exists,' I told the class, 'you don't need a depth psychologist to point it out. It isn't always a dusty subterranean tomb full of mystical statues and sacred icons. It isn't just the burned fragments of parchment entombed in the Mediterranean mud off Alexandria. It isn't the ancient gods swimming under the surface of contemporary religions, and occasionally sucking down a human sacrifice or two.

'Often enough,' I said, 'it's simply a kind of malignant stupidity. You can find that anywhere.'

The best examples of that malignant stupidity are the near-successful suicide attempts Western Civilisation has made since the start of the 20th century. The First World War comes immediately to mind, and the Second.

So does the nuclear arms race, which occurred after, not before, Hiroshima and Nagasaki had shown what nuclear weapons could do to cities and the people in them.

The genocides come to mind: the American Indians, the Armenians, the Ukrainians, the Jews, the Bosnians and Rwandans, Congolese and Syrians, to name just a few – where some essential awareness in the killers had to fade away before they could destroy the lives of their neighbours.

Even in places where wars and genocides succeeded, one consequence was the permanent loss of consciousness for the perpetrators. When it comes to murder, the crazed courtroom faces of the old dictators tell us that the conscious mind cannot face what the unconscious has done.

Malignant stupidity is implicit in the growth of capitalism, at least if you graph capitalism's projected expansion over the next fifty years, and imagine what will be left of a soft-lit green and gold world after it's run through capitalism's rendering machine.

It's in continued population growth – the explosion followed by the crash, followed by further crashes as the environmental destruction of the original overshoot hits home.

When you view the unconscious like this, it's like glimpsing a moronic reptile grin peeking from a broken window in a ruined city.

Back on the deck, I ask Jamie, 'How long have we got? I've got friends with grandchildren.'

'How long?' he asks. "Time really isn't an issue from my standpoint. Maybe you should go back over what you said in class.'

'What was that?'

'Check your notes,' he says. It takes awhile, but in my office I find a faded folder labeled *The Unconscious as Literature*, and take it back to the deck.

The unconscious isn't aware of time, I read. *If it's a place, time doesn't exist there. If it's just the non-neocortex parts of the brain, it has no more concept of the future than cows calmly walking up the curved ramp of a well-designed slaughterhouse.*

That brings up a dilemma. Is time something that exists independently of consciousness, and consciousness just makes us aware of it? Or is time just the illusory byproduct of consciousness and every minute of this class is a false division of eternity? Theologians, theoretical physicists and the people in charge of nuclear waste storage would like to know.

'Pretty insensitive words to say to a kid who's been told he's got weeks to live,' Jamie says. 'I had nightmares about that curved ramp in the slaughterhouse.'

VI

Also on the class reading list was an essay from the 1980s, by the plain-spoken Jungian esssayist Michael Ventura, titled 'Cities of the Psyche'. The essay notes an uncanny similarity between satellite photos of cities and photomicrographs of computer chips.

Ventura suggests that in each case we're building structures that turn the unconscious into flesh —yet we don't consciously design chip architecture, we just figure out ways to make a chip smaller, cooler, more compact, and better at manipulating ones and zeroes.

We don't consciously design cities, either, we just try to satisfy criteria for real-estate profits, master bathrooms, commuting, lawn care, pizza delivery, backyard barbeques, cocktail parties, baseball and soccer and adultery.

Ignore the scale, and computer chips and cities look identical. Ventura has the idea that something is using humanity as a tool to create itself. Something beyond the human is making us build structures too small or too big for humans. It can't all be due to body-image disorders.

So you can have the pyramids of the Egyptians and the Maya, the sudden appearance of echoing cathedrals all over Medieval Europe, inhumanly rational skyscrapers with inhumanly empty nighttime streets below them, great monolithic hospitals named after saints, missiles named after gods, musical instruments shaped like women, guns shaped like men – toward the end of *The Unconscious as Literature* we had begun compulsively identifying the unconscious components of our world, making connections like Michael Ventura makes connections, seeing the campus clock-tower as homage to the god Priapus, our college cheerleaders as sacrificial virgins to the same god, and the Middle East-bound ROTC students as sacrifices to the nasty old god Moloch, who promised ever increasing prosperity in exchange for the children.

One connection I deliberately didn't mention was that our hospitals were an incarnation of disease, and disease was an incarnation of the unconscious. In that direction lay a scapegoat-the-cancer-victim madness, and I didn't want to go there while Jamie was still with us.

But Jamie went there for me, saying, 'Having cancer is like having all the scenarios of the unconscious hit you at once.

'The basement of my hospital is an underworld,' he went on. 'In it is a refrigerator, and that refrigerator contains a petri dish full of cancer cells with my name on it. It's one of a bunch of petri dishes. They're all full of live cancer cells. All of them have names on them. Some of those names belong to people who are dead.

'What's in that petri dish will outlive me. It will outlive you, too, unless some lab tech forgets to feed it, or feeds it too much and it outgrows the refrigerator.'

Despite his total lack of hair and the divots in his physique, Jamie was a good-looking guy. He had girlfriends that he had met in the hospital or at Camp Rainbow Gold, the wilderness camp for cancer survivors where he volunteered every summer. The problem was that Jamie's girlfriends all had their own petri dishes in the refrigerator.

Jamie said, "Cancer shares a brain with you. You're never sure if it has you or you have it, and in the end it doesn't matter because you're tied together and you can never figure out where you end and it begins.

'And cancer is the reptile brain doing what it has to do. The way my life is now, I feel my neocortex is an afterthought tacked onto the other parts of my brain. My doctors tell me that my cancer is cells that should have died out before I was born but didn't. They were here first.'

VII

No doubt Jamie's cancer cells, even now sitting in the hospital basement along with all their fellow campers – Camp Rainbow Gold for the neoplastic set – consider themselves perfectly conscious and focused on the business at hand, which is to divide and grow and keep dividing as long as somebody's supplying the nutrients. They might look unconscious from the outside, but they wouldn't agree, thank you very much. They've got all the consciousness they need for the job they're doing.

We have to look at things from Jamie's standpoint to realise his cancer cells are insensate entities in an underground refrigerator, kept alive for research purposes, dependent on human beings who are trying to figure out ways to kill them *in vivo*. Conscious humans are part of the unconscious of every cancer cell. They're not aware of us, although they might have an uneasy feeling that something out there beyond the petri dish doesn't mean them well.

How do you get from the standpoint of a cancer cell to the standpoint of Jamie, who has a different understanding of the petri dish and its nutrients?

VIII

We have become a depressingly aged and unfulfilled civilisation, as civilisations go. Time has caught up with us. Where once we were full of promise and intelligence and a lust for life, we are now sticking with the known and the comfortable. In financial terms, we're living on interest rather than producing useful products. In agricultural terms, we're eating the seed corn. In ecological terms, we're parasitic.

We've begun to feed on our young, not just because of college loans and home mortgages and the military campaigns of a tottering empire, but also in our fascination with their quick energy and apparent immortality. We hold them close, keep them living at home far beyond the time when they should leave, limit their independence and their options and network them into jobs and careers that will have them living within our own diminished horizons before they're thirty.

The metaphorical diagnosis has come down, and it's not benign. Things we didn't want to think about or couldn't think about, or even couldn't have known about at the time we made fatal decisions – the frenzied waste of fuels, the impossibility of endless growth in a finite world, inequality that dehumanises both rich and poor, the religions that promote big families – all these things should have died out before our civilisation was born, and maybe if we had devoted enough energy to becoming aware of their consequences they would have. But we stayed unconscious, and they stayed alive, and they became an integral part of who we are. Now they and their metastases are killing the world.

They haven't killed it yet. But give them time. Better yet, take time away and put eternity in its place.

In eternity, all exponential curves go vertical. Unintended consequences dance out into the open for everyone to see. The slow increase of greenhouse gasses in our atmosphere becomes lethal when seen through the dark lens of eternity.

So does the best-designed nuclear power plant. In eternity, nuclear waste ponds dry up and explode into radioactive flame.

In eternity, longevity becomes dementia. Oilfields go dry. Oceans warm and turn to acid. Effective antibiotics last for geological nanoseconds. Pesticides become nutrients for weeds. Species wink into and out of existence. Deserts crawl across continents. Corporations are born, grow, wither and die. Countries do the same. Civilisations do the same. Suns expand into the orbits of their planets.

As Jamie might say, there are as many endings as beginnings in the big scheme of things.

IX

I deliver Jamie a cup of coffee on the deck. We make the sad discovery that he's not solid enough to pick it up. 'That's OK,' I say. 'I'll drink it.'

'I miss coffee,' he says.

'I'd die without it,' I say.

'Not funny.'

'Tell me,' I say. 'Does consciousness exist after death?'

'Sure,' he says. 'But there's argument over whether or not it exists before death.'

X

If, by the end of *The Unconscious as Literature*, we had no idea of what the unconscious was, my students at least knew how to recognise the not-so-subtle signs of it in their lives. They began to look critically at their relationships, and say, 'I know what you're really saying when you say that.' There were breakups, prefaced by the words, 'What Jamie Would Say is that you're a total jerk.'

Other students, who previously couldn't stand each other, fell in love – because they could finally see what each other was really saying when they said what they said, and even if it wasn't nice, it was true, and once you're aware of the unconscious, true is always preferable to nice. Not that they don't sometimes go together.

Students began to do their own laundry rather than face the unconscious choices attendant to taking it home for Mom to wash on weekends. They began to talk to their grandparents about their parents, learning embarrassing family secrets. They began to look critically at their professors, even when I discouraged that sort of thing.

In short, because they gained a perspective on the unconscious, the students began to show the sort of aware maturity that small liberal-arts colleges point to when they try to justify tuition increases. Some of this new perspective was due to me, some of it was due to the readings, but much of it was due to Jamie and his conscious ability to say what he really meant at the same time he was saying it.

Over my career, most of my good classes did have someone who was dying in them, whether anybody consciously knew it or not. When those people did die, their death gave their lives meaning, in a way hard to articulate while they still had a future in this world. Whether it was an actual physical death or a *Lolita*-style death didn't matter as much as the fact that in each of those long-ago classrooms, one of us transformed death from an abstract idea into something that gave daily life heft and substance.

These days, it's the impending death of our civilisation that is endowing what remains of our lives with heft and substance, and the classroom is the Earth. It's time to play one more game of What Would Jamie Say.

'Here's what I'd say,' says Jamie, trying and failing again to pick up his coffee. 'It's the things that humans do know, rather than the things they don't know, that are determining the fate of the planet.'

'You mean we're not lemmings, unconsciously diving off a cliff?'

'You're lemmings consciously going off a cliff,' he says.

'You used to tell us that one way of approaching the unconscious was to realise it was out to get us. But the unconscious is just the world. If you want to make any part of it conscious, take it seriously and look at it honestly. That's all consciousness is: taking things seriously and not lying to yourself.'

'If you think of it that way,' he says, 'consciousness doesn't come from the self. It comes from outside the self. It always was outside of the self, in plain sight. The self just has to stop defending against it.'

Then he mutters, 'I really miss coffee,' and disappears into the under-world, or back into my unconscious.

I'm all alone in the sunshine, and it's getting warm out here and I'm without sunscreen in a world where the ozone layer is still delicate despite our best efforts, and that question I was going to ask Jamie – about how consciousness can exist in a place where time doesn't – will have to wait.

I pick up Jamie's full coffee cup, which suddenly looks like a legacy, and look out at the mountains. In the past, they've had snow on them this time of year, but it melted fast this spring. Only small patches of white mark the north-slope gullies. Most of the trees below them are beetle-killed. There's smoke from seven hundred thousand acres of nearby burning forest in the air, and a layer of wood ash has drifted onto the deck.

It's not too much to say that humans have become the Humbert Humbert of species, and the young green world we feed on has become our Lolita.

But in my small part of that world, autumn will come, bringing a para-doxical purity. The air will clear and cool and if the fire hasn't made it to the deck and the house, I'll still look out on a beautiful, if blackened horizon.

At that moment, I'll know that having Jamie in my class made me a better professor. I'll know that Lucretia, the smiling keeper of Jamie's narcotics supply, is more real when she's got a patient in pain. I'll know that Lolita would have become a slattern even without Humbert Humbert, because it's in the nature of nymphettes to age and thicken. I'll know that gazing out on a burnt-over world brings beauty into being.

I can ask these questions only by forgetting what I already know. There is much to be said for asking questions that can be answered by looking carefully and without prejudice at the world, even if the answers then look like common sense.

We could actively observe as a civilisation and discover that we know more about our problems than we think we do. We might even find out enough to save ourselves from the doom-ridden future that is haunting us.

But we've reserved our questions of policy for the unconscious, forgetting that the unconscious – besides being a storage locker for malignant stupidity – is an endless and eternal and infinitely complex set of circumstances that takes things only as seriously as it has to. It doesn't have to answer our questions. Unlike consciousness, it doesn't have to care and it doesn't have to grieve. Common sense says that it's stopped taking humanity and our little sparks of consciousness seriously.

Against that indifference, I can only sit and talk with Jamie on these delicate, orange-lit mornings on my deck, with coffee. I realise we won't save civilisation. But we might save what's left of Jamie until fire season comes around again, and that will be something in the face of nothing, and that is enough for now.

Small Gods

MARTIN SHAW

Issue 7, Spring 2015

We hear it everywhere these days. Time for a new story. Some enthusiastic sweep of narrative that becomes, overnight, the myth of our times. A container for all this ecological trouble, this peak-oil business, this malaise of numbness that seems to shroud even the most privileged. A new story. Just the one. That simple. Painless. Everything solved. Lovely and neat.

So, here's my first moment of rashness: I suggest the stories we need turned up, right on time, about five thousand years ago. But they're not simple, neat or painless. This mantric urge for a new story is actually the tourniquet for a less articulated desire: to behold the Earth-actually-speaking-through-words again, something far more potent than a shiny, never contemplated agenda. As things stand, I don't believe we will get a story worth hearing until we witness a culture broken open by its own consequence.

No matter how unique we may consider our own era, I think that that these old tales – fairy, folk tales and myths – contain much of the paradox we face in these stormriven times. And what's more they have no distinct author, are not wiggled from the penned agenda of one brain-boggled individual, but have passed through the breath of a countless number of oral storytellers.

Second moment of rashness: the reason for the generational purchase of these tales is that the richest of them contain not just – as is widely purported – the most succulent portions of the human imagination, but a moment when our innate capacity to consume – lovers, forests, oceans, animals, ideas – was drawn into the immense thinking of the Earth itself, what aboriginal teachers call Wild Land Dreaming. We met something mighty. We didn't just dream our carefully individuated thoughts – We. Got. Dreamt. We let go of the reins. Any old Gaelic storyteller would roll their eyes, stomp their boot and vigorously jab a tobacco-browned finger toward the soil if there was a moment's question of a story's origination.

In a time when the Earth is skewered by our very hands, could it not be the deepest ingredient of the stories we need is that they contain not just reflection on, but the dreaming of a sensual, reflective, troubled being, whilst we erect our shanty-cultures on its great thatch of fur and bone?

It is a great insult to the archaic cultures of this world to suggest that myth is a construct of humans shivering fearfully under a lightning storm, or gazing at a corpse and reasoning a supernatural narrative. That implies a baseline of anxiety, not relationship. Or that anxiety *is* the primary relationship. It places full creative impetus on the human, not the sensate energies that surround and move through them, it shuts down the notion of a dialogue worth happening, it shuts down that big old word animism. Maybe they knew something we have forgotten.

Two routes towards the cultivation of that very dreaming were through wilderness initiation and, by illumination of the beautiful suffering it engendered, a crafting of it into story to the waiting community. Old village life knew that the quickest way to a deep societal crack-up was to negate relationship to what stood outside its gates. Storytellers weren't always benign figures, dumping sugary allegories into children's mouths, they were edge characters, prophetic emissaries. More in common with magicians. As loose with the tongue of a wolf as with a twinkly fireside anecdote. These initiations facing the rustle-roar of the autumn oaks or grey speared salmon had banged their eloquence up against a wider canopy of sound, still visible on the splayed hide of their language.

Part of a storyteller's very apprenticeship was to be caught up in a vaster scrum of interaction, not just attempting to squat atop the denizens of the woods. To this day, wilderness fasting disables our capacity to devour in the way the West seems so fond of: in the most wonderful way I can describe, we get devoured.

The big, unpalatable issue is the fact that these kind of initiations have always involved submission. For a while you are not the sole master of your destiny, but in the unruly presence of something vaster. You may have to get used to spending a little time on one knee. May have to bend your head.

Without a degree of submission, healing, ironically, cannot enter. It is not us in our remote, individuated state that engenders true health, but soberly labouring towards a purpose and stance in the world that is far more than our own ambitions, even our fervent desire to 'feel better'.

So, I claim that the stories are here. And they include all these difficult conditions. That's the price tag. This is not in any way to claim redundancy to modern literature, but simply to hold up the notion of living myth.

The Seven Coats

CHARLOTTE DU CANN

Issue 6, Autumn 2014

'From the Great Above she opened her ear to the Great Below.'

I am taking off my red coat. In its pockets are seeds, rose hips, bus tickets, notes from meetings. The coat has mud on its woollen sleeves where I have dug festival ditches and community gardens, stains where I have poured tea in church halls and slept in protest tents, where I have chopped wood in my garden, a badge on each lapel that says 'We are the 99%' and another that declares freedom for Palestine. *We can turn the ship around,* I have been writing these last six years, *we can do it ourselves. We can repair, resolve, remember, restore, re-imagine the world we see before us falling apart.*

I stand in the corridor, with the six coats upon their pegs, lined up like so many books on a library shelf: my life laid out in sequence. I wanted to write how it is when you leave the coat on a hook, pulled by a line that was written four thousand years ago.

I wanted to tell you about the first yellow coat, as I walked beside my mother down Queensway, London, how it determines all the others. It's made of primrose Harris Tweed, signalling that I come from a certain class of beings who live this city. This is my first moment of consciousness. *I am me!* I declare and in this moment break away from her.

My mother walks onward past the sawdust floors of butchers and the cool leafy interiors of grocers. It is the end of the '50s. I am a small light in a darkened city. This feeling I realise does not come from my mother, or my father who is working in the law courts of the city, defending small murderers and thieves. I know, even though I do not yet have the words, that this existential moment is stronger, more alluring, more meaningful than anything I am surrounded by.

To be free, to awaken, to be your true self, to know the secrets of life you have to let go first of your mother's hand. To live is to know how to die. But when you have died, you also need to know how to be reborn. And to recognise that moment when it comes.

When Inanna tricked her father Enki of the Me that conferred on her the powers of her office the greatest she held was the gift of discernment.

FASHION My adult coat was not always red, or second hand. Once it was tangerine and new and caught the eye of my friend Alexander in Rome.

'Why have you got the hook outside of your coat?' he asked.

'It's a fashion detail,' I said. 'It means the coat is by Jean Paul Gaultier. It's his signature.'

Alexander laughs. We are on the Spanish Steps and my friend the sem-inarian quizzes me with all the force of his Jesuit education. I don't tell him this is the most expensive coat I will ever buy, or why that deep orange embroidered frockcoat was the only colour and shape to be wearing that season. Or that why in spite of all my learning I am writing about men who design beautiful things.

'Who is he?' he says.

The question you have no answer for, that holds you to account, is the one that shifts everything.

CLOAK Once the coat was a grey cloak with a scarlet lining with my name stitched in its collar: blue to signify my house, Ridley, named after the Christian saint. Inside its deep inner pocket there is a battered copy of *Ulysses,* a book I will silently devour, while the rest of the chapel will pray to a god who spent three days in 'Hell' before rising to the sky realms. The institution has taught me to sing psalms, recite Shakespearean metre, pro-nounce French verse, and, in moments of disobedience, read Joycean prose without a full stop.

I have learned from these texts that the true power in writing lies not in clever argument, but in listening: but only from the last do I learn its greatest trick of all, which is to break the rules.

FUR When I was twenty, I broke the rules of all my class and education and went to Belfast to be with my first love and he gave me a coat made of soft grey rabbit skins. He had worn it when he was in a rock band. We stood on the Ards peninsula and watched a hundred wild swans land on the black sea. It was the middle of the 1970s, and all my encounters were ventures in uncharted territory. From my lovers I discovered how it is live in the industrial north, in South Bronx, to be a Jew, to be ashamed of poverty, to be a policeman, to be sent to the madhouse, to prison, to fight with god – subjects never mentioned in my father's house.

'How come you are the hero in everything you write?' asked the man I did not sleep with.

I did not know. I was experiencing life by proxy.

BROWN When I go on the road to experience life for real, I will wear a honey-brown car coat that once had a belt when it swung in the Dover Street shop, alongside cedar drawers of soft silky shirts from Tibet. My sister gave it to me one freezing winter's night in New York and afterwards we went out like furry twins to catch a cab and to eat Moroccan and drink large glasses of pinot grigio.

The alpaca coat will serve as a blanket in the cold mountain nights in the Andes and Sierra Madre. I don't fly anymore, or eat in restaurants. When I think of New York now I remember the tramp on Broadway who told you: *you have somethin' golden inside there, in your brain; y'all take care of it, y' unnerstan'?*

BLACK 'I like to see you smiling there,' said my father as he lay dying, and the summer storm raged outside the hospital window. In my hand I was holding a raven feather, now buttonholed in a small black frock I found in a thrift store on our last road trip to Utah.

I wanted to tell you how it was when we arrived in Zion Canyon that spring, how it was when my father's spirit roared into the night, the stories held within the fabric of each of these coats, but each time I go there I run out of words and a small quiff of terror runs through my veins.

I am standing in this corridor facing the coats and realise they are no longer my store of material: not these childhood nostalgias, these *Bildungs-romans*, these young rebellious love stories, these glossy magazine articles, these poems about birds and ancestors, treatises on plant medicine, not even the latest narratives about collaboration and downshift.

What next now that everything is written, now there are no hooks left?

The Line

In the introduction to her retelling of the Inanna cycle Diana Wolkstein writes of her first encounter with the Sumerian scholar, Samuel Noah Kramer. Kramer had been working with the four-thousand-year-old inscriptions for fifty years, a cycle of myths and hymns she would describe as 'tender, erotic, shocking, and compassionate – the world's first love story that was recorded and written down'.

From the Great Above she set her mind to the Great Below.

'What exactly does mind mean?' she asked.

'Ear,' Kramer said.

'Ear?'

'Yes, the word for ear and wisdom in Sumerian are the same. But mind is what is meant.'

'But – could I say "ear"?'

'Well you could.'

'Is it *opened* her ear or *set* her ear?'

'Set. Set her ear, like a donkey that sets its ear to a particular sound.'

As Kramer spoke, Wolkstein recalls, a shiver ran through her.

'When taken literally, the text itself announces the story's direction. From the Great Above the goddess opened (set) her ear, her receptor for wisdom, to the Great Below.'

'The Descent of Inanna' is the fourth and final myth in the quartet, and the four together are understood to be the cycle of a complete human being – specifically a female being. This final part records how the Queen of Heaven and Earth goes into the Underworld, where she is killed by its Queen, her sister Ereshkigel, and then is restored to life.

Inanna has to go through seven gates before she gets to her dark sister's throne room. At each gate she has to give up one of her Me, the attributes of civilisation, from her crown to her breechcloth – all seven seats of her physical and material power. She enters the *kur*, the Underworld, to know the secrets of rebirth housed there, which are not the physical attributes of the middle earth but belong entirely in another dimension.

You shiver because you know you can't follow the words of her myth in your mind. You follow her track the way dancers hand down their choreography through time: by imitation.

The Myth and the Story

The myth is not the story. The story is extrinsic. I walk out, fight dragons, lose myself in the forest. I return, get married, live in a castle, inherit the kingdom. I do this, then I do this, then I do this, then I hang up my coat on the back of the door and tell you a story. You listen to my tale, gripped by adventure. It fits into the ordinary world we know. Our lives are built around these stories with their happy or sad endings. We are rewarded or punished, the good triumph, the bad die, or do a far, far better thing and suffer both fates.

But the myth is not this. It demands we open our ears to another wavelength. It is complex, non-linear, and runs alongside the story of our middle-earth lives, with its clawed feet in the Underworld and its beaky head in the sky realms. It doesn't fit with what we see around us. It lives in caves and out in the desert wind, and sometimes looms up in the city darkness and tells us to take care of something inside us that we cannot see with our everyday eyes.

When the story loses its sense, the myth emerges like the bones beneath the soil. It promises something that makes sense beyond the endings we predict, yet leaves us puzzled by its inscriptions on stone and clay, with its bird heads, its masks and painted bodies. With the goddess who rides on the back of a lion, who is conquered and then transformed.

The myth is intrinsic. It works from the inside out, looping back on itself, and lives in all time. In myths, like our dreams, there are savage things that don't make sense. You cut off the heads of people who seem to be giving you direction or asking for help. You eat the things you should not, and open the box you should not. You are married to your father and your brother and your son. You are a strange heroine. Discernment is your greatest gift. Curiosity and a thirst for knowledge pulls you where angels fear to go.

Angels don't lose their clothes, and in the Underworld you lose everything. The clothes are the least of it.

I am standing naked, before the hook and my sister's wrath. The myth will kill me and put my body on the hook for three days, which is the statutory amount of time a soul stays in the Underworld before it returns to the sky realm. My ascent will involve complicated deals with sky fathers and loyal servants, betrayals and praise, and someone I love who will take my place. Nobody goes into the Underworld and returns. Except you who break the rules.

The ways of the Underworld are perfect. The ways of Heaven are perfect. I am imperfect and incomplete. Like all earth creatures I bring change by undergoing change. As a people we can change the law, but only through our own journey which demands we give everything away that up to that point has conferred power upon us.

Civilisation tells us we should stay still, be perfect and never change. It gives us coloured coats to wear and says by these outer forms you shall be known. But this is not the life that illuminates our being. You go into the Underworld to find that out the hard way. It takes off the layers one by one, peels them, all your worldly colours, until you stand stripped in the strange twilight of the Underworld, infused by its lamps of asphodel.

Mostly you go to meet your sister, whom you have been told is furious with you. Somewhere buried in this myth from Sumer is a key about the future. And for weeks now I have been waiting for it to appear. The first known piece of writing was written by a woman c. 2200 BC in praise of this being – who was not a mother goddess, but embodied the morning and evening star – and her myth of descent is the first of the 'mysteries' to emerge from the city cultures we call civilisation.

It is hard to imagine a world shaped by such a descent, because we live in a world framed by monotheistic gods, who sacrifice their sons to war and Empire, and sentence their daughters to servitude. You have to go beyond millennia of saints and masters and sages into the strife-torn deserts of modern Iraq to find where Inanna first held sway, before she became by association the whore of Babylon, her alchemical moves reduced to a strip-tease of coloured veils, performed for a bored tourist in Istanbul.

Embedded in her myth is a way to go beyond civilisation's impasse. Because the life ordered by the Underworld is not the life ordered by Empire: it has another structure and practice entirely. As modern people we like to hold the myth philosophically, culturally, psychologically at arm's length. What we fear is to walk in its tracks, lose control over our lives. We do not like to question our existence at every turn. So we toy with the mythos in our minds, at the end of our typing fingers. Ereshkigel, we say, is our shadow, and become small professors in the arts of deities and griffins. This means that, we say, with our breasts puffed up like chickens. It's about numbers, and cycles of planting and growing, the seven planets, seven colours of the rainbow, seven chakras. Inanna is a fragment from the matriarchal era. She is Venus who appears as the morning star, disappears under the Earth, and reappears in the evening.

But information is not the myth. Myths are *enacted,* dramaturgical, pro-tean, existential. You allow the myth to be played out through your being, suffer its effects consciously. The meaning and the expansion it brings happen inside of you, wordlessly. When you stand by the hook, you are scriptless. Libraries disappear, all your smart lingo of Eng Lit and fashion and philosophy. You are in the place without words. The words take you here and then abandon you.

Writers are born with the kind of memory that calls them to go through the gates of the kur. They remember, not just for themselves, but on behalf of the people: we have to undergo change, or we are not people and the Earth is not the Earth. When we make our moves the edifices tumble down, the institutions crack, illusions dissolve like mist.

It comes to me in this moment that I have run out of the storyline. I don't know the ending to my own story, or that of anyone around me. And maybe this life isn't a story anymore. Maybe it's something else. The future stands before me like an empty quarter, like the desert road edged with sunflowers, like the twilight in the garden after the rain. I take a deep breath. *I am here,* I say and step forward.

The hook holds what you most fear, which in my case is meaninglessness. The void hits you like a mallet and you tremble. You break apart like a seed pod. Collapse happens inwardly and suddenly.

At the moment Inanna is killed by her sister, Ereshkigel begins her labour. When her servant Ninshubar goes to heaven to ask Inanna's fathers to rescue her from the hook, the first two refuse. Then the third, Enki the god of wisdom, creates two beings made from the clay under his fingernails who slip into the Underworld unnoticed and assist Ereshkigel to give birth by sympathising with her pain and glorifying her greatness.

Oh, oh, oh, my inside, oh, oh, oh, my outside!

Inanna goes into the Underworld because she knows her sister has something more powerful than any of her Me. That's what pulls her, that's what pulls us, thousands of years later, caught by the first line. We are hooked on that moment.

Some of us have been so hooked on that moment we forgot what we went down there to find in the first place.

Leaving the City Inside

The story of civilisation tells us we will be rewarded if we toe the line: but though some may receive a moment of glory, or own a fine house or dine on meals that slip extravagantly past our lips, none of this will give us kinship with the beasts, or our fellows, or return us whence we came. None will tell us what we need to undergo to become real people – which is to say people who value life on earth.

The myth tells you if you give everything to life, the Earth will give you everything your heart desires: which if we are writers, means knowledge is given to us – a lineage that stretches back through time, to this moment when our words were first inscribed in clay. That is why we go to the Underworld and face the hook, even at the risk of losing those words that have kept us safe all these years. All those poems and articles, adjectives, and smart lines. All those narratives.

The writer is the one who remembers the myth and keeps telling it to the people. Nothing happens for the better unless we let go and change our forms.

The ways of the Underworld are perfect, Inanna. Do not question them.

What is hard for our duality-driven minds to comprehend is that Inanna and Ereshkigel are the same being, that to turn the ship around we have to follow her mythic track. Rebirth takes place in the Underworld, and in order to reclaim, remember, re-imagine, we have enter to its domain.

And we absolutely don't want to go down there. We want to stay in our cosy colour-supplement lives and cling to our ideas of happy families and romantic love, our knowledge of buildings and history, our Shakespearean quotations. We long to keep our shirts perfectly ironed in cedar drawers, to repeat the epithets that fall from the lips of holy men in robes.

Who am I without these coats of class and institution?
Who am I without my work?
Who am I without my new found community?

When Inanna returns to the Great Above the person who has not mourned her departure is made to take her place. Her consort, the shepherd Dumuzi, who is also Tammuz and Adonis and Dionysus, and all dying and resurrecting ivy-wreathed gods of the ancient world, and further down the line, the sacrificed man on a cross who does not remember her name. Whose books tell us we don't have to go there, because he did it all for us.

The rebirth we seek does not happen without our descent. The world becomes flatter, uglier and unkinder, determined by the unconscious mass, the untempered leader, the foolish woman, the words that do not set their ear to the Great Below. Venus, the embodiment of love, beauty and a fair fight, steps into the arena to bring new life. She doesn't do that by chanting a new mantra or changing her shopping habits, she does that by grabbing you by the throat and pulling you towards everything you have so far refused to see or hear. She takes you towards the unspoken, the missing information in every transaction, each time you have jumped the consequence and refused to hear the beast or child cry out, your sister trapped in a factory a thousand leagues away.

The unconscious snarls back, rages and rants, complains, resents our every intrusion. It is not polite, or reasonable, or forgiving. You have to withstand its every humiliation: inside yourself and outside amongst the people you love and fear. We think to know the facts is enough, that good behaviour is enough, that to write of our wounds and sorrows is enough. But it is not enough.

To let go of earthly power is a real thing. To be conscious within the realms of unconsciousness is a real thing. To face your raging sister, to move out of the cycle of history, to liberate yourself from your line, to have empathy for the man, for the child, for the tree, for the fish and the barbarian, these are real things. Not to give up, even when you have given up and the world has turned its back on you.

To die before you die is the core tenet of all the mystery cycles that emerged in the early city states before the father gods took command. It has been a task undertaken by writers in the civilisations that followed – content that we labour conveniently in the Underworld as volunteers and substitutes to carry their shadow and suffer on their behalf.

But Inanna's myth does not end there.

Exodus

It is the moment I hang up the red coat. The moment I expect the hook and find none.

I am on the beach on a warm blue July morning. There is one day a year like this, and today it is here. The sea shimmers and stretches out before us at low tide, and the breeze carries the dusty scent of marram and sea holly. In the sea the currents move around the sandbar, this way and that, and tumble me into the foam. Every time I put my feet down the sand moves too and small fish who lived buried in the seabed. Everything is moving. I am laughing, tossed by the waves. This is how it is on the tip of the future, as you look at the sun on the horizon, as you look at the empty page and don't know what to write anymore.

I wanted to tell you what it is like when you have done your time in the Underworld, the moment that delivers you into a vast unmapped space, and frees you from the past that has been howling and pawing your coat it seems for centuries. I wanted to say how it was all worth it, though I am left naked on a beach, bookless, featherless, empty-pocketed. Because at this moment I want to be nowhere else but here, with the future unwritten before me. Because the golden feeling I had in the core of my self when I was two years old is still with me at fifty-eight, and keeping loyal to that awakeness is what I steer by, what I trust more than anything I see falling apart around me, and I know I am not alone in that. And mostly because I remember what my sister told me before I left the city:

'You have been the anchor, you have kept this house together, you have absolved our father's guilt, buried our mother with honour, held our hand, listened to us, grieved with us, written our story, now it is my turn.'

I put my feet on the firm wet sand, on the shoreline, on this beautiful day. *We are here*, I say.

Wet Sage and Horse Shit

ERIC ROBERTSON

Issue 4, Summer 2013

The great nigger-head war of 1972 began the day Uncle Earl's oldest daughter dumped melted 'I Can't Believe It's Not Butter' down the back of the khaki short pants worn by Deverle's second oldest boy, David. The Christensen family reunion had been put off for a month due to some family disturbances, so tensions ran hot between Deverle and his older brother, Earl. Their usual campsite, the high Uinta mountain meadow at the Smith and Morehouse Campground, was taken by another family. The site they had to settle for was farther from the stream and overrun with hundreds of Black Angus. When Deverle arrived with all his kids, Uncle Earl's twin boys were caught in a bloody fistfight. One of them forgot to hobble the other's horse when he put it out to graze. The free horse ended up with its leg caught in the ventilation shaft of an old silver mine. Earl had to pry the boys apart. Threw one of them into a horse trough full of pond water. The other one tripped and fell into fresh cow turd.

Earl had the rest of his boys tramp down the tall grass and clear sagebrush so the family's five trailers could park in the traditional schooner circle. His girls cut vegetables and peeled potatoes for the Dutch oven. They got into a fight over why one of them bought the fake butter. The youngest of Earl's kids set out to collect an arsenal of nigger-heads, which is a violation of the agreed-upon rules of engagement. Both teams of Christensen, the Earls and the Deverles, were to be present before harvesting could begin. The Earls had gained an unfair advantage. The kids quickly hid their cache of dark blossoms when they saw dust kicked up down the road.

Deverle was anxious as he approached the camp. His older brother knew how sensitive he could be, so Earl made sure he and his wife were the first to run to the truck and greet him. Deverle's kids threw open the spring-loaded door on the truck's camper and piled out the back. One of them unlocked the door on the small second-hand trailer that was being pulled behind. Deverle got the dented caravan in a trade for a month of water rights. The rest of the young Deverles poured out and ran into the arms of Aunt and Uncle Earl.

Deverle sat in the cab of the truck. David sat next to him. Quiet. Deverle expected the news about what had happened to reach over to Bear River and Earl's boys were going to give David hell for it. Deverle looked over and thought about a couple things he could to say to his son to help prepare him. He took a minute. Didn't say either one of them and slowly stepped out of the truck. David sat there and knew the longer he stayed in the cab the more likely it was that everyone could tell that something was wrong. But then, like rusty clockwork, through the cracked windshield splattered with bugs, David saw the Twins. They stalked the truck like a couple of wolves and swung a lasso. David smiled. They hadn't found out. Everything was going to be all right.

I don't know if you can sense this or not. I'm putting off telling you the whole story. Not because I'm uncomfortable with this kid, but because I'm a bit ignorant. This kid hadn't figured out where his body ended and other boys' bodies began. If I were to tell you how many times and where and with whom this kid got naked, there'd be a scandal. And that scandal would get pinned to the back of his head and it'd stick to him the rest of his life and I wouldn't want that. We do have rules for nakedness and young boys with changing bodies and I don't want to undermine that either. But as much as I believe in adhering to this agreed-upon decency, I hate half-stories. So I'm going to try to explain to you what I think it meant for this kid to want to always see his twin cousins naked. I'm no expert on kids who are that way, but I'll do my best.

Deverle raised a wrestler. A damn good one. The kid was rarely pinned. He held the regional record for escapes and beat the Hostetler boy in under a minute, something nobody thought was possible. But word got around, true or not, that David was making lewd comments to his opponents during matches. And that there was inappropriate touching as well. It was half a dozen kids who came forward, even kids who had beaten him, so it wasn't just the sore losers. What settled the matter in the minds of most folks was how often David got an erection during his matches. And his was one he couldn't hide.

The championship match was bruising. But he won it. After it was over David was exhausted. He stayed on his knees, on the mat, head down. The referee held his opponent's hand. The whole gym waited. So David stood up. There it was. Not a damn thing he could do about it. And it was one of the only nights his father was able to come. Deverle ran the late shift at the dairy and it was a rare treat for him to watch his son wrestle.

His opponent called him an awful name. Mothers and cheerleaders pretended not to notice. There were a lot of whispers and giggles. David stood there and let it be. The ref raised his hand as the victor. There was scattered booing. His teammates ignored him. His coach gave him an indifferent pat on the back and David walked out to the showers.

Deverle watched the rest of the boys wrestle and sat quiet as everyone left the gym. David never came out to meet him at the door of the locker room so Deverle put on his coat and sat in his truck in the empty parking lot. A thick mat of frost covered the windshield. A cold fog herded all the light into a single triangle that hovered over the truck. He understood what lay ahead for this kid. There wasn't much he could do. He couldn't stop February, March, and April from coming.

In May David pitched a tent up Blacksmith Fork. He took a camping stove and his rescued Mustang and changed the tyre on his brother's dirt bike so he could use it to get around. He left home to live in a tent because his father didn't know what to say. David lived in that tent until it was time for the family reunion.

Deverle was surprised at how happy his brother was to see him. The Twins, like every year before, set up the bucking barrel for David. Nothing seemed out of place. The news of what happened hadn't made it over the mountain yet.

This year the Twins made the bucking barrel from an old oilcan, used to collect rainwater from a busted cabin. With heavy chains they brought from Earl's machine shop, they suspended the can from four giant posts still standing from an old cattle ramp attached to the empty corral. Before the Twins could drag David out of the cab of the truck, he jumped out. This summer he didn't put up a fight.

'Oh shit,' one of them said.

'Uncle Deverle this one's gonna hurt. Keep your truck running,' said the other.

David's legs that year were longer and thicker. They fit around the barrel. He dug his heels into the rusted metal. He pulled on his own fitted work gloves. With his hands, opened and closed, he made two snug leather fists. The Twins strapped the rope around the barrel and over his hand as tightly as they could. David had only minutes before his fingers went numb.

'Here we go cowboy.'

Deverle stood back. He was afraid. His heart had broke so many times for this kid. It just felt like it was meant to happen again. The Twins had hardly ever bucked David off the barrel. Michael, Deverle's oldest boy, wanted nothing to do with rodeoin'. His middle boys, Paul and Alma, were too reckless. They weren't careful enough with dangerous things, so they could never develop technique. His girls weren't meant to be up on there and his babies weren't even old enough to ride bikes yet. So he had David. They tried to work through some things before the trip. They went to the Bishop together to talk about the situation. Afterwards, the Bishop confessed to Deverle in private that he had no experience with kids that were that way.

So they called the Brethren in Salt Lake and they sent a man out that David could talk to. Nobody knows what was said. David signed some papers and the man left in a black car.

Deverle shuffled the worn baseball cap to the back of his head and looked down at his feet. He kicked the dirt. A small shadow moved over his work boots. Swallows were out. There weren't many that year. It had been real dry. But at least the aspen had bounced back from the bark cankers. The sky was clear after a night of rain. The sage was wet. Deverle could smell the horses. David stayed on the barrel. That frustrated Earl's older boys so they dumped a bucket of muddy water on him as he jumped off. He couldn't free his hand and they left him tied to the barrel. Deverle's girls loaded up on nigger-heads and shot them from the wrist rockets Deverle gave them for Christmas. The Earls took cover. Uncle Earl laughed and ducked the barrage to get to the barrel. He helped David loosen the rope.

'That's a good ride son.'

'Thank you, sir.'

Earl extended his giant hand. David smiled big and took in heavy breaths of air. His hand felt broken, but he quickly grabbed his uncle's and gave it a hearty shake.

I'm going to stop right here. I realise I've started telling you a story you've already heard before. It feels to me like you think you know what's coming. Like you've got things figured out. This must be a true story. The author's writing about himself. You've figured this is my confession.

Some of these characters are me. Some of them aren't. Some of them are relatives or friends or people I've heard of. The summer I describe is real. The tamped grass and diverted streams. The old corrals and knotted fence posts. Some of these are people I wish I could have been. People I wish I could have been closer to. You've heard this kind of beginning to a thousand different stories. It's happening right now. In high mountain meadows, groups of the gathered generations of families collate and rearrange their Western legacies. But how prepared are you for a story like this to proceed? A kid who is that way, on a landscape made for other men. If finishing this story makes you uneasy, do it the simple, old-fashioned way.

First, pick a clown. Any clown. The man in a dress. The tortured father. The muscle-bound hustler addicted to methamphetamine. The better-dressed half of a newly-wed couple throwing a hundred grand at a female surrogate to take up their mixed semen and make them a legacy.

Now pick the tragedy. A man raises six kids with a depressed wife for the appetite of a voracious religion. A young wrestler hangs himself in a barn, over cow shit and rusting metal. A ranch hand is dragged behind a horse by his testicles through an open field of prickly pear. Pick any one of the myriad ways you know how people at your margins are harangued and killed. Any one will do. You'll read that novel, won't you? Give that film an award? Name a plant after me?

Now think of my final tableau. Where do you put me? Up there, on that highest peak with grizzlies and bighorn sheep? Can you imagine me herding cattle with dogs and spindly grandfathers? Do you see me in coal mines, driving railroad spikes, or splitting rails? Can you see me on the backs of bison? Tell me what I have to do to stake my claim on this same piece of mighty ground. And what of my family? How do you imagine them, having bred the likes of me?

Up from underground came my father and grandfather. Up from underground a crack opened, filled with copper, then scoured into an open wound. Up from underground came the violent cost of electricity, an illuminated world's fair, street lamps, and hotel chandeliers. Miles of coiled copper scratched out of the earth inch by inch in lengths that drove men mad. Men who bought and beat women. Unconscious women who abandoned their sons. The picture of such a man, my grandfather, a man who did violent, awful things hangs above my window, framed, next to a picture drawn by Dr. Seuss. Is that tough enough? That I honour a publicity photo from 1933 of Blackie Robertson, a prizefighter, a romantic mongrel of the violent American West. A part of the world that chews on these leathery, worn-out notions of manhood like dogs at a rawhide. If I stake that claim on a character that drank more than he ate, who knew only vulgarity and wore it as a hardened necessity, if I stake that claim, frame it and square it off, then will you let me give names to things, to colourful plants and birds, to rock formations and cliff art? If I say I am like the men that have come before me can I name a river or a waterfall or an Indian tribe? Can I then help contribute to the mythos of an American West? Or would you still have me dead in the attempt?

Think of a different story for me. A mind that has no concern for offspring. What does such a primitive brain think? When I stand in a field without the infinitude of generation, without that sprawling family tree rooted in my groin where do my thoughts go? What do I see?

I see the beauty of my own body as it lives in a present tense. My sex explodes without concern for generation or increase. Without increase there are no storehouses. Without increase there are no fences. Without increase there are no creeks diverted, pooled and haemorrhaging in clay ditches or leaking through metal troughs. My possessions perish every evening, used up with the setting sun. Without increase there is only wet sage and horse shit.

I stand in a field, eyes on my feet, barefoot, craggy toenails and cracked skin. I will never pray. I am now my own centre. My skin, my only boundary. Without me to define you, your centre is broken open. Like a cracked egg on hot, pitted cast iron, your centre bleeds and spreads. Your borders are coming apart. Your margins are abandoning you.

Once you stood in your straight-edged fields, eyes cast to a vacant sky. At your margins you dug holes, strung jagged wire between poles of pine and poplar. You married after the first bleed. You mapped and drew lines that cleaved mountain peaks in half, parted wetlands, and furrowed the bloody backs of Africans and the soft spaces between women's legs. At the margins you sliced through the faces and chests of boys strapped to split rail fences when you found out they were that way. You stood there, in your field, surrounded by degrees and latitudes. Lined the perimeters with your favourite plants, carelessly named them for the colours of sunshine, the skirts of beautiful women, and African slaves. This is how you've fenced yourself in.

I have no land to protect, for children, for paternity, for posterity. In my field I am but a present passing moment. A brief spark. Not given to definition or category or ill-fitting names. I do not reckon with square and compass. My earth is pocked and irregular. I leave nothing behind, because before me there was nothing. In a field with no margins I stare into a pond at my feet. I see the refracted image of increase. The beauty of one eye for one eye. One birth for one passing away. I see the beauty of sex, of liquid light, of indulgence. I see expenditure without return. I see my own miraculous brevity. I see the beautiful dead.

David, Deverle's that-way son never got beaten up or down. He helped his fellow Deverles win the nigger-head war. On his second attempt he got bucked off the barrel on purpose which endeared him to the Twins for his toughness. That act of bravery helped him convince the Twins to swim naked in the hidden pools on Beaver Creek.

He never did assault young athletes in football locker rooms. Didn't marry other men in conciliatory, watered-down churches or make his increase from the rented wombs of desperate women.

David lives in fields under borrowed tents. He counts winter eagles in salt marshes. He markets backyard produce. He is naked often, early and unexpectedly. And he has learned the proper name for the nigger-head. *Rudbeckia occidentalis*. Western coneflower.

FIVE

Humans are not the point and purpose of the planet. Our art will begin with the attempt to step outside the human bubble. By careful attention, we will re-engage with the non-human world.

EMILY LAURENS – Remembrance Day for Lost Species 2014 – Memorial to the Passenger Pigeon – *Llangrannog Beach, Wales – Issue 7. Spring 2015*

We need to find new ways to commemorate the passing of places, ecosystems and species. 100 years ago and 4,000 miles away, the passenger pigeon became extinct. From billions to none in the blink of an eye. With that in mind, I went to where the civilised, farmed, human-dominated land meets the wild untameable ocean to draw passenger pigeons with my friends in that luminal space. Within those fragments of rock that sea and time have crumbled to near dust, I trace their shapes with my garden rake. In the blond brown sands I draw a small flock of pigeons, like shadows passing overhead. And then I watch them disappear. *Photograph by Keely Clarke*

On the Centenary of the Death of Martha, the Last Passenger Pigeon

PERSEPHONE PEARL

Issue 7, Spring 2015

Aldo Leopold, 1947, writing after the unveiling of a statue dedicated to the memory of the last Wisconsin passenger pigeon, shot in September 1899
Men still live who in their youth remember the pigeons. Trees still live who in their youth were shaken by a living wind. But a decade hence, only the oldest oaks will remember, and at long last only the hills will know. There will always be pigeons in books and in museums – but these are effigies and images, dead to all hardships and to all delights. Book pigeons cannot dive out of a cloud to make the deer run for cover, or clap their wings in thunderous applause at mast-laden woods. Book pigeons cannot breakfast on new-mown wheat in Minnesota, and dine on blue-berries in Canada. They know no urge of seasons, no kiss of sun, no lash of wind and weather. They live forever by not living at all.

Etta Wilson, resident of Petosky, Michigan, and eyewitness to the events in the woods in May 1878
Day and night the horrible business continues. Bird lime covers everything and lies deep on the ground. Pots burning sulphur vomit their lethal fumes here and there, suffocating the birds.

Gnomes in the forms of men wearing old, tattered clothing, heads covered with burlap and feet encased in rubber boots, go about with sticks and clubs knocking down the birds' nests, while others are chopping down trees and breaking off the over-laden limbs to gather the squabs.

Pigs have been let loose in the colony to fatten on the fallen birds, and they add their squeals to the general clamour when stepped on or kicked out of the way.

All the while, the high, cackling notes of the terrified pigeons, a bit husky and hesitant as though short of breath, combine into a peculiar roar unlike any other known sound, which can be heard at least a mile away.

Of the countless thousands of birds bruised, broken and fallen, comparatively few can be salvaged – yet wagon-loads are being driven out in an almost unbroken procession, leaving the ground still covered with living, dying, dead and rotting birds. An inferno where the pigeons had builded their Eden.

1857 Ohio State Senate Select Committee report

The Passenger Pigeon needs no protection. Wonderfully prolific, having the vast forests of the North as its breeding grounds, traveling hundreds of miles in search of food, it is here today and elsewhere tomorrow, and no ordinary destruction can lessen them, or be missed from the myriads that are yearly produced.

Once upon a time

An old story tells of a commonwealth of birds, where there were countless different birds of all shapes, sizes, temperaments, appetites. Their vast principality spread from ocean to ocean, from snowy mountains in the north to desert in the south, with birds perfectly adapted for every space.

Wandering the seas and coasts were loons, grebes, albatrosses, fulmars, shearwaters, storm petrels, tropicbirds, pelicans, boobies, gannets, cormorants, darters, frigates, jaegers, gulls, terns, skimmers and auks. Diving in the lakes and bays were herons, bitterns, storks, ibises, spoonbills, flamingos, swans, geese and ducks. Birds of prey roamed the skies: kites, hawks, eagles, harriers, ospreys, caracaras, falcons and vultures.

Grouse, ptarmigan, quails and turkeys nested on the heaths and uplands, while cranes, limpkins, rails and gallinules dwelt in the marshes. Coots, oystercatchers, stilts, avocets, plovers, sandpipers and phalaropes trod the shores. Owls and nightjars hunted in the dark. Parrots showed off their dazzling plumage. Cuckoos laid their eggs in others' nests. Kingfishers, woodpeckers, tyrant flycatchers, larks, swallows, jays, magpies, crows, titmice and nuthatches ate caterpillars in the forests and meadows. Dippers, wrens, mockingbirds, thrashers, thrushes, gnatcatchers, kinglets, pipits, waxwings, and shrikes all sang their hearts out. Vireos and warblers were known as the sprites of the woodlands. Meadowlarks, blackbirds, orioles, tanagers and finches lived in jubilant flocks. Swifts, hummingbirds and pigeons were superb aerialists.

Eventually, humans arrived too. The birds watched them, and saw how they hunted, how they sang songs, how they raised their children. A conference was called, to see what should be done. After long deliberation, the birds decided to welcome the humans to their domain, and discussed who should offer what gift. The Carolina parakeets and the Ivory-billed wood-peckers offered their plumage. The Bachman's warblers offered their songs. Great auks offered their soft down and their glistening fat. And then the birds looked around to see who would offer themselves as food. All eyes fell on the passenger pigeons, of whom there were so many. And the passenger pigeons said yes, there are enough of us: some of us will offer our bodies to the humans as food, to make them welcome, to share our beautiful world with them.

So a single white passenger pigeon flew down from the conference to a Seneca camp by the side of the Allegheny River. She landed on the shoulder of the oldest person there, and told him what the birds had decided. I don't know what he said in reply.

Dispatches from Bastar

NARENDRA

Issue 4, Summer 2013

Bastar is a region of India inhabited almost entirely by Adivasi, or tribal people. My association with Bastar goes back to 1979 when I first went there in the course of my work, and in later wanderings with some Gandhians. A little later I undertook field studies in a part of Bastar known as Abujhmad, studying the Adivasi worldview and how it interacted with the modern world.

Spread across 4,000 square kilometres of in places impenetrable vegetation, and with a population of around 13,000, Abujhmad is a very remote and 'undeveloped' part of Bastar. It had had little or no contact with the outside world when I first arrived. Some interior villages had never known the impact of the wheel. The small community lived on food gathering and hunting, with shifting cultivation as a supplement. It had neither trade, industry, occupation nor other modern apparatus. But neither was there hunger, starvation, beggary or lingering disease.

For many years I have been writing and sending out 'dispatches' from this tribal area, with the aim of helping modern people to understand the Adivasi worldview and to help Bastar to retain its vigour and vitality. These are two of them.

I. The Language of 'Issues'

It was late evening. The birds were returning home, while children of the village played, raising clouds of dust noiselessly. They played as noiselessly as the birds overhead, or the movements in the forest. Despite the existence of plenty of sounds, one often wonders at the absence of noise and dissonance in the deep interiors of Bastar; at the absence of the disagreeable. The wild, untamed resilience of deep forests is a space for meditation and self-reflection.

Ever uncomfortable with 'issues', I was trying hard to explain to my friend Nureti what climate change is about. Such difficulties had been run into earlier, too, with many I have spoken with on other issues. How does one explain an 'issue' to an Adivasi or, in the first place, how does one make an issue?! How does one explain issues to people when, in either of the languages, ethos and resilience are amiss or at variance?

I do not know, but in my experience it was probably in the 1970s that language began taking its strident turns. Like capital, language too began to be modulated by the few. As an instance, when the word 'environment' arrived sometime in the 1980s, it was difficult to explain to my father. He was an educated man. That was about the first time I discovered the emerging dissonance between me and my parents. Issues have replaced languages; they have guile and deception.

So, I struggled to explain climate change to Nureti as I had heard and read of it. Increased carbon emissions, carbon footprints, climate not as a social but economic issue, carbon markets, carbon sinks, changing crop patterns, rising sea levels, impending disasters and catastrophes etc. As I spoke, there was a hesitation. I was doing to Nureti what I had done with my father about 30 years ago.

The hesitation also stemmed from the changed focus and idiom of global debate on climate change as having been hijacked from a social-political plane of ordinary human collectives – their experience, wisdom, nurturing institutions and agencies – by the exclusive political-scientific sections which continue to hegemonise the globe. How was I to explain climate change to one who does not wish to step beyond his village; for him the beyond is a pursuit futile and foolish, a transgression, un-silent and shrill. He has a persuasion different from mine; and, in my experience, very healthy, reassuring and honourable.

After listening patiently, Nureti urged, 'Do not spread falsehood, it shortens the life of Earth. When our gods and goddesses were living they had vitality to shape the world and do things good for us. Now they are stones. The patient stone, however, speaks if we heed it speak. What you say are words. Your word has taken away the vitality and the promise; but like our gods it is not living either. Now vitality and promise have left your living word, too. That which is without promise is evil and dwells in darkness, causing depravity. It is fickle, it keeps making and breaking. It leads astray the passing wayfarer; it may give some joy but there is more grief.

The *marin tan podela* (a plant which causes temporary amnesia if brushed against) betrays one into psychosis and amnesia, for then one remembers neither home nor hearth, nor the ways to them. It leads one into the unenduring. Your word is of the netherworld; soulless. It is born of a spectre, mortal and perishable; here today, gone tomorrow. Don't spread such word. It is falsehood and shortens the life of Earth.'

As against the language of healing and correction, how often and unwittingly we celebrate the language of malady and impairment. Such is the power of 'issues', their language and currency. Such language is corroding; it precipitates distress and suffering amongst the global majority that doesn't speak thus. In its wake, people suffer loss and grieve over ways of life; rituals of celebration, healing and correction lose significance too. For such majorities they are invasive discourses. How immensely, and deeply, have we damaged such people through the brute power so readily at our disposal; and how irredeemably are we endangering ourselves now that they too are beginning to speak this language.

Has the marintan podela betrayed us into psychosis and amnesia? For now we seem to remember neither home nor hearth, nor the ways to them.

II. Of Toilets and Ancestors

Every now and then the government announces a plethora of welfare plans for Adivasis. They pertain to forest, land, water, poverty, mining, malnutrition, mortality, housing, education, agriculture, livelihood, debt and bondage, among other things. Political parties and civil society groups clamour for expedient delivery of such welfare. For the year 2004-05 the government of Chhattisgarh allocated 93.68 billion rupees towards 25 developmental plans. All of them speak of Adivasi 'welfare' at village and regional levels.

A folk tale in Bastar mythology deals with a monster who recurrently blocked waters from returning to ancient habitats and parched the human community to gradual extinction. The tale has quite possibly become real today and threatens the Adivasi community with decadence and extinction. Given the impact of welfare plans and schemes, Adivasis do not look like 'stakeholders' in their own welfare. More often than not, the nature and content of plans reflect systemic prerogatives rather than Adivasi prerogatives. It is a direct affront to their sovereignty and wellbeing as a people, and diminishes the community's character, the single most important Adivasi resource.

Given the plans and schemes, the traditional modes of designing and constructing houses have changed. An Adivasi remarked to me on the unsettling irony of such welfare:

'Traditionally, construction of a house began with creating and dedicating a space to the ancestors. But there has now come a scheme [from the World Bank] which provides a subsidy if you build a toilet in the house.

Subsidy is granted after the toilet has been built. With the bulk of land taken away, the plots are small. There is no space for toilets. If you want to utilise the scheme you have to first make a toilet. So we have had to let go of the space for ancestors. In that space we now make a toilet, first and foremost; construction begins with the toilet.

We have lost our thatch huts; we have lost even our ample drinking water; land and forest; we have lost our grass, too. Our grass grew tall; now there are some kind of bushes that destroy the grass. Our cows fall ill more often. It needs money to treat them. We are losing our ancestors, too; this is the last blow. We belong to none; none belong to us. We are adrift now. Things have swung.

We always lived with our ancestors in our houses. They gave us blessings and favours. We fulfilled our filial duties of food or propitiation towards them. Depending on their earthly dispositions they brought us disease or wellbeing, strife or peace, and guided us through life in our dreams or possession. They protected us. When we recovered from illness or harm, we made sacrifices; gave thanks with glad hearts because we knew they may save us on another occasion. That is how we stayed together. We were a family of living and dead relatives. We believed in the dead, not death.

I may not know my ancients, their names or stories of their lives. But I know my mother, for who could know her better than me! Who could know better than her children how much she loved us! We remember her kindness while she was living. She will treat us in the same way now that she is dead. These schemes have brought us grief and lamentation. Because of toilets we have lost our dead. When we visit the deeps of our forests, cross the swollen rivers or visit another village they will not be with us. Governments cannot love like parents.

I am an Adivasi, and this is my house. I have been living in this house for long. I built it according to certain norms of the community that we revere. My observance of these norms kept me in good kinship with my community, the land and forest, the here and hereafter. And now that is being displaced. There is a new way coming up which has nothing of the community, nothing of the influences that affect my and my kin's lives, and through which we made more of ourselves. Whatever I needed to build a house – some land, some wood, some mud – were all available. Then came cement and concrete. Though it had to be purchased, but it was available. Nobody from outside could tell me how I should make my house or what I should make in my house...

Our houses and villages were healthy and clean. The waters were sweet. Our children and animals roamed free and did not have many diseases.

We communed freely with the living and dead. Now we defecate where our ancestors fed. We cook, eat and empty our bowels under the same roof. This is what we shall bequeath our children. We have alike forsaken our ancestors and children. Everything has been given away; even our self-esteem. Who, then, needs the jal, jungle and zameen [water, forest, land] There is nothing Adivasi in these schemes or in the various plans, acts, ordinances and legalities that are made and fought for in our name.'

The worldview of commerce, the market and technology – in various garbs and measures as governments, political parties, corporations and social-political groups – is the increasingly absolute engine of change and governance. It is sweeping everything before it. In Bastar, the only countervailing force seems to be the assortment of Maoist groups, active and powerful in the area. But they, in effect, are nurtured and sustained by the same impulse, the same garb and measures of life and living; the impulse of dishonour, disengagement and marginalisation. The Adivasi is hard-pressed.

The Dispatch is based on a conversation sometime in 2006 with an Adivasi of Tondamarka village.

How Wolves Change Rivers

KIM MOORE

Issue 6, Autumn 2015

By singing to the moon, when the beavers move in, by the growing of trees, when the soil resists the rain, when the sky rubs its belly on the leaves, by singing to the wind, by killing the deer, by moving them on from the valleys, by the birds coming back to the trees, by singing to the water, with the return of the fish, with the great ambition of beavers, with the return of bears moving across the land like dry ships, by an abundance of berries, by the bear reaching and pulling down branches, by the green coming back, by the green coming back, by the steadiness of soil, by the deer leaving the valley and the gorges, by the aspen growing, by the cottonwood growing, by the willow growing, by the songbirds singing to the trees, by the beavers coming back to love the trees, by the absence of coyotes and the abundance of rabbits, by the bald eagle and the raven who arrive to minister to the dead, by the glove of a weasel and the burn of the fox, by the gathering of pools, the holding together of the river bank by the trees, by the river finding its spine once again.

Rampant Rainbows and the Blackened Sun

CARLA STANG

Issue 6, Autumn 2014

The Mehinaku Indians of Brazil believe that the current state of affairs in the area where they live is destroying what might be called the very 'essences' of their landscape. Here I want to explore this belief and the methodological and philosophical ramifications of taking people seriously. The urgency of the environmental scenario calls on anthropologists to reconfigure their relationships to what they study. If instead of using theories that distance them from the people they work with, they in fact believe them, that is, accept what people like the Mehinaku are saying on their terms, I will suggest that anthropologists and others are in this way provided with precious thousands-of-years-old knowledge for how to live with ongoing environmental devastation.

I

The Mehinaku, with whom I do fieldwork, are an Amazonian people who live in the Upper Xingu of Brazil. This area constitutes part of what is called the Xingu National Park, a large government-administered indigenous reserve. To the Mehinaku everything in this landscape is made from things called the *yeya*. Not only each human body and creature but every object in their sur-roundings is a copy or imitation generated from an 'original' called a yeya.

These yeya, or 'original things', still exist today at the bottom of waters and in the forest. I was told that there is literally a yeya for every kind of thing that exists, including even the things of white people (*os brancos*[1]) such as cars and watches! There are yeyas for the birds and fish, ceramic pots and houses, jaguars and feather headdresses. It was first explained to me that a yeya is an exact, true-to-colour miniature version of its copies, from palm-size to forearm length. One of my informants, Arako, told me that his uncle Sepai, who is known as a very powerful shaman of the Yawala-piti community, kept two yeyas of an armadillo (*ukalu*) secretly hidden in his house. When they were children, Arako's cousin showed these to him.

Inside Sepai's straw mat-roll for tobacco and other shamanic things, Arako saw the yeya. It looked exactly like an armadillo, except it was very tiny, the size of his hand. It was extremely delicate and moving, shivering, alive.

These yeya are the immortal, 'true' versions of all things that surround human beings in their world. Before the sun first shone, before the first human beings, there existed little people called the *Yerepëhë*. These tiny kindly folk were at home in the dark so with the first sunshine they hid in masks. These masks, or 'First-Forms' as they might be called, are the yeya, and are understood to exist now as the essential bodily form of spirits.

These yeya of the spirits are the archetypal forms from which inferior copies, the things of the world, are issued. The way the Mehinaku express this 'firstness' is with the Portuguese word for 'true' — *verdadeiro*, as a translation of the Mehinaku *washë*. For example, a fish yeya is called *Kupatë-washë*, which means 'fish-true', distinct from an ordinary fish, which is called *Kupatë-ënai*, 'fish-cape/clothing'. These 'True-Fish', the *Kupatëwashë*, live in the deeps of the waters, and are also the stars of the shining river in the sky, the Milky Way. They are the 'chiefs' (*chefes*[2]— *amunão*) that make regular fish, *Kupatëënai*, for people to catch and eat. The latter 'fish-copies', the replicas of the 'first', or 'chief' fish, were explained to me as 'fish-shirts' (*camisas de peixe*). They have an ordinary, not particularly satisfying taste and are tougher in texture. On the other hand, on the rare occasion that a 'True-Fish' is caught, these are found to be fatter and softer, with a flavour that is extraordinarily delicious. These special fish can only be captured using white people's poison, not with an arrow or hook as they know to avoid these.

How the replica-skins actually come forth from the yeya is unclear, but there is definitely some sense of organic reproduction, with the *apapanye-yeya* understood to relate to its *ënai*-copies 'as mother to children', with an apapanye-yeya often referred to as, for example, 'Mother of the Fish'.

In all of this there is a sense of material connection between the yeya and their ënai, the latter derived from and thus containing a trace of the former's body, and associated with this, an emotional relationship of the caring parent-to-child kind.

II

This sense of the paradigmatic constitution of the Mehinaku world is understood by them to be under threat from the impact of the world beyond it. The progressive changes in the reserve have been of quite a different nature to what has gone on in other parts of South America. Unlike other indigenous peoples of the continent, the Mehinaku and other Xinguano people's experiences of contact with the outside world have been largely indirect, through disease epidemics, the use of manufactured goods and interaction with restricted visitors from outside, rather than by direct day-to-day economic and political subjugation. Because of the types of contact they have experienced, and because of the way the communities have responded to successive crises by reconstitution and integration, in many ways the Xinguanos have thus far been able to maintain their way of life to a far greater degree than most other Amazonian peoples.

Having said that, far from being the protected paradise the Brazilian government purports it to be, since its founding in 1961, the Xingu National Park has been subjected to continuous invasions and ongoing destruction. These have included rapid deforestation (especially for soybean production, as well as logging, corn farming and cattle ranching), pollution of water and contamination of the fish that are the mainstay of the Xinguano diet. The most devastating threat is the Brazilian government's highly contested building and running of hydroelectric dams in the region during the last decade. According to the Xinguano peoples themselves, the situation has now become critical, bringing their leaders together in July 2007 to declare the crisis in an open letter to the Brazilian nation.

This is an overview of the deterioration of environmental conditions in the Xingu basin. I have described it in Western terms: 'environmental degradation'; contamination of clean water with agro-toxins; the chopping down of oxygen-yielding trees; the threat of extinction of species. In Mehinaku terms this state of affairs concerns the harm and obliteration of the yeya. As we discussed, the yeya archetypes still exist shimmering and alive in the landscape. As the farmers and loggers cut away at the edge of the forest the yeya that live in those particular parts of the forest are destroyed or retreat in anger. Likewise the poisoning of the water by 'os brancos' threatens the yeya of the rivers, streams and lakes. As I mentioned above, the yeya of fish, which appear as big especially sweet-tasting fish, can only be killed using white people's poison. Arako, a Mehinaku man, explained to me that this is in fact why the number of fish are decreasing: that is, white people's poison has killed too many 'True-Fish' and those that are left angrily cease to make their replica-fish and do not allow themselves to be caught. As the Mehinaku see it, the spirits are enraged and even annihilated altogether. When the spirits are angry they refuse to issue from their bodies the imitations that are the things of the world. That is, the very 'building blocks' of the world are ceasing to work as they should and in some cases are being completely destroyed. Existence as the Mehinaku know it is being deformed, unformed and could eventually cease, as the archetypes of existence stop generating the substance of the world and disappear altogether.

III

According to the Mehinaku, white people are destroying the world unwittingly. Principally because they lack understanding of how to properly be and act, generally and in ritual.

The Mehinaku cultivate a particular kind of consciousness throughout the day. This way of being is called *awitsiri* or 'beautiful way' and is mostly marked by its quality of 'care'. This is the 'state of mind' that nurtures all things towards which its attention is turned. This careful way of thinking, feeling and doing is evident in the manner in which people generally strive to conduct themselves. Some individuals are seen to be particularly successful at achieving this active state, and after a while it was apparent even to me, the calm and painstaking way certain people attended to even the smallest task.

An important part of this 'care' is the quality of attention given. Once, a young man called Maiawai sought to teach me the importance of this way of doing things. I was in a hurry, pulling in frustration on a zip that seemed completely stuck. He laughed and stopped me, and motioned to watch as he sat and with slow, calm deliberation attended to the task. He knew I was time-pressed and still he instructed me, *'devagar, Carla, sempre devagar'*[3], ... slowly, Carla, always slowly'. The careful nurturing concentration is a form of the principle of childrearing *(paparitsa)* that is extended to all matters in the world. It is the opposite of the destructive effects of uncontrolled desire, as in *awitsiri*-consciousness one's desires are restrained into caring form. For example, the desire to open the zip is not allowed to go out of control as it was near to doing in me (that probably would have resulted in breaking the zip). Careful loving attentiveness to the task preserves the integrity of the thing (Maiawai closing the bag, all parts of it still intact). The importance of this intensive kind of 'paying attention' in fact has the tone of a 'spiritual' injunction wherein *Kamë*, the Sun, chastises a lack of concentration by giving the negligent person an injury. I found this out one day when Arako chided me for my scratched hands, cut from grating manioc, asking me whether I had been thinking about something else while I was working, and laughing and scolding me that *Kamë* had punished me for this with those injuries.

One of the nurturing and preserving aspects of this 'caring' consciousness is a certain 'lightness'. Note the laughter in both cases above. There is
also a certain lightness in the way Mehinaku people look after their children.
They are cared for in a far less controlling manner than in the West. They
are allowed to go and do what they want from the time they are old enough
to wander off, only casually telling their mother where they are going if they
are going far. They work from a young age but only if they choose to. Literally,
when they are young they are only held with the lightest touch and as they
get older, from the time they can walk properly, they are hardly touched at
all. There is an expansion of this delicate conduct into the quality of consciousness. There is a sense that, as in the treatment of children, to be too
attached in thinking and feeling, to grip too strongly, may be damaging to the
integrity of any entity one relates to, as well as the relations between entities.

Another aspect of the awitsiri consciousness is that the careful attention
involved intertwines one utterly with all aspects of the worlds around. To
at every moment devote complete attention to the things and creatures one
encounters is to be joined in strong relationship to all those entities.

This special awareness is brought to the special things the Mehinaku do
to mend relationships with the apapanye spirits of the plants and animals
and other aspects of their world, which anthropologists would call 'rituals'.
How the Mehinaku fix problems in their landscape, heal illness, find lost
things would be better called a kind of technology, something that is real,
something that works according to a systemic logic, not a leap of faith or a
simple belief in magic. Here again we find ourselves taking people seriously
instead of documenting a quaint belief. There is not space for me to fully
describe this Mehinaku 'physics' so I will just give a brief outline. The basis
of rituals for the apapanye is the association between the representations
made by the Mehinaku and the original apapanye. The link is material
in a similar way to how the apapanye-yeya is substantially linked to its
ënai-copies as described earlier. The representation made by the Mehinaku
draws the spirit-entity close according to a principle similar to magnetism.

It is evident here how clearly and manifestly the Mehinaku understand
ritual, the forces at work known though invisible. The apapanye-spirit is
pulled in by a physical link, where it stays in the vicinity of the ritualist. The
ritualist does not somehow 'become' the apapanye in some kind of mystical union, the apapanye is understood to be simply 'looking on' as it stays
invisibly close. What is more, the presence of the apapanye is so distinctly
understood that it was once drawn for me in a diagram as a shaded region
completely surrounding the figure of the person. This aura contains the

material associations (smells, feelings, recent deeds) of the ritualist, which if offensive to the associations contained in the apapanye's own aura, will repel the spirit-entity; the overlapping and interaction of the constituents of two substantial 'clouds', more chemical reaction than mystical possession.

IV

White people, the Mehinaku say, do not live in an awitsiri way. They do not attend to the world around them with careful, light, nurturing attention that preserves the forms of things. They are not caring or attentive or able to enter into proper communication with the forces they share their world with. They are exploitative in their unrestrained desire, thus neglecting or utterly destroying relationships with other kinds of entities and often the things and beings themselves. As we have seen in cases such as that of white man's poison killing the True-Fish, the 'things' being destroyed include even the 'First Things', the yeya, and when these are offended, damaged and destroyed it threatens the continuity of the aspects of existence of which they are the source.

These are only a few aspects of the Mehinaku sense of dangerous changes in their world. I will briefly mention one other. The Mehinaku have a sense of consensus realities, of how groups of the same soul-kind with their collective perceiving, uphold a certain manifestation of reality; how the Mehinaku and neighbouring Xinguano groups are similarly human in perspective and thus manifest a reality of peaceful continuity of the world in its proper form. Groups of souls of a wild kind such as other Indian groups and white people, and spirits, manifest realities of unrestrained desire that cause disintegration of the world and ultimately chaos. The invasion of other destructive realities can be in the form of the invasion of spirits, particularly at the time of eclipses, when the unnatural darkness causes dissolution of the forms of the world. The other way this happens is by the descent of the rainbow snake to Earth where he wreaks havoc. Like the collapse of the circle arc of the rainbow to Earth, the encroachment of white people is also perceived in terms of circles; as the collapse of the concentric boundaries of the worlds on the ground. The outermost ring – the wildness of white people – is penetrating inwards and if this continues there might be a complete implosion, the world of forms dissolving, leaving a world of chaos, a return to the primordial darkness of spirits without form, desire utterly unrestrained.

So how do the Mehinaku live with this sense of a world dissolving into chaos? They keep going. Because of their sense of consensual reality, they know their responsibility, that is, how critical it is that they maintain a sense of calm and caring, of relationship to and hence preservation of the structures of their world. All of this, in the face of an aggressive war of cosmologies that they are aware of and we are mostly not. Their consensus and thus manifest reality against ours. Outwardly they fight (mostly indirectly) where they can – they have lobbied their government for their lands more effectively perhaps than any other indigenous people of South America – and laugh when dignity is all that's left. For them the world has always had threats of dissolution, by rampant rainbows and the blackened sun. So they have always had to be vigilant, they are well practised at it, and perhaps this is why they might teach us white people a thing or two, we who have always been so sure of our world. They live with some of our paraphernalia, our t-shirts and plastic containers. When these things enter their realm, they use them up, quickly, then they burn them. They hold to their way, which they call *anaki*.

V

The very real possibility of ongoing, unstoppable ecological devastation can beg the general question: 'who would have good answers to all of this?' The way that I see it is that there are no easy answers, and if we are looking for more fruitful ways of living in a world in flux we do not need to start scratching around making rudimentary things up. Anthropologists encounter myriad of such lived cosmologies, and these are not skeletal outlines but elaborate thousands-of-years-old, tried-and-tested lived forms of knowledge; it is only a matter of attending to them, of taking them seriously. If the bankruptcy of the hegemonic Western worldview has brought us teetering to the edge, it is anthropology's special disciplinary position that may offer some of the alternatives.

In a sense the positive aspect of the apocalypse has begun, that which is the literal meaning of the word, from its Greek *apokalupsis*, to uncover, reveal (*apo-'un-'* + *kaluptein* 'to cover'). What has been uncovered at this time is the barrenness of the dominant worldviews, the rampant greed and heedless destruction of consumer capitalism. And in anthropology the tendency to simply mine other people's experiences to use in elaborating Western theorisation. In most of the old ways of anthropology, people's lives, wherever they are lived or of whatever character they are, simply become grist for the theoretical mill, interpreted through Western models that are based on other Western models, in a continuing fetishisation of Western conceptions. We are thus kept in a loop talking to ourselves, getting ever further away from the experiences of the people we started with.

The alternative is to stay with the experiences, 'returning to the stream of life whence all the meaning of the words and theories came from', as William James (1947:106) put it. Instead of breaking up the details of people's existence and putting them to the service of extrinsic explanations, one may attempt the opposite: to take Western language and its categories (the explanatory media mostly at hand) and make *them* the grist for the analytical mill, mixing and shaping them in order to make a description of people's experience. This is basically what I have sought to do in this piece of writing.

Let us return to the question, 'Who would have good answers to all of this?' I suggest that making such painstaking renderings as described above, of different cultural experiences, allow people to enter into other worldviews, so as to see how they might be taken on (the 'how' is an open question) as more hopeful ways of being in the world. For example, what if we – anthropologists and others – attempt to suspend disbelief and entertain in our own minds the possibility that for example, the Mehinaku are simply right and that there are in fact yeya, the mostly invisible archetypal presences; that there might in fact be aspects of the world that we are not aware of and in our ignorance are neglecting and even harming, and thus damaging the landscape in as yet untold ways? And having taken people seriously on this point – of the possible existence of numinous but specific unseen forces – we are provided by people such as the Mehinaku, with thorough knowledge of how to best live with such forces, how to live the awitsiri way, with keen awareness of the environment, lightly and with great care. In fact in their rituals the Mehinaku have an entire system of creating relationships with the yeya, of not only not destroying but nurturing the essential and generative aspects of their landscape and helping mend what has been broken.

Ecologists create models of the interconnectivity of thriving ecological systems. Famous scientists devise theories about how the planet might be more than what it has seemed to us since the so-called Age of Reason, such as Lovelock's idea of the planet as Gaia, a living being. And yet these theorists do not have to start from scratch. If they look over the disciplinary fence – as scientists such as Newton did in the past – they will find anthropology's riches containing wondrously elaborate versions of such ideas. Similarly, philosophers and social psychologists might have a lot to gain by attending to anthropological material, like the Mehinaku's sense of consensual realities, which shows the great care humans need to devote to the quality of our perspectives. Let us find ways to take other people like the Mehinaku seriously: perhaps join them in their special skilful laughter that protects the essential shapes of the world, or their helplessness they know how to turn to resourcefulness as things are falling apart.

References

1-3 These words are Portuguese. Most other words italicised in the first place are in the Mehinaku language.

For a more detailed description of how Mehinaku 'ritual' works *see* Stang (2009). Stang, C. (2009) *A Walk to the River in Amazonia: Ordinary Reality for the Mehinaku Indians*. New York, Oxford: Berghahn Books.

Other references include: Gow, P. (1991) *Of Mixed Blood: Kinship and History in Peruvian Amazonia*. Oxford: Clarendon Press; James, W. (1947) *Essays in Radical Empiricism*. New York. London. Toronto: Longmans, Green and Co; Lovelock, J. (2000) *Gaia: A New Look at Life on Earth*. Oxford: Oxford University Press.

Squirrel

DAVID SCHUMAN

Issue 6, Autumn 2014

It was in the grass. He thought it was dead. It seemed as if it had been there for a while, like the grass had grown around its body. But it wasn't dead. It quivered as he drew near.

'Go climb a tree,' he said to it.

He nudged it with the edge of his flip-flop. It sprung up as if shocked and landed on the top of his foot. As he tried to shake it off, it bit. Then it fell from his foot and ran across the yard. It pressed itself against the fence.

He reached down to touch the bite and then thought better of it. It was more of a pinch.

He went inside.

'A squirrel just bit me,' he announced.

'What?' his wife said. She sounded sceptical and accusatory at once.

Dolly looked up from the table.

'Maybe it has rabies,' Dolly said.

His daughter was a sullen and frightened girl. She perked up at any suggestion that the world wasn't as safe as her parents assured her it was.

His wife dried her hands on a towel. 'Maybe you should call someone,' she said.

'Who?'

'Animal control,' said Dolly. She was the authority on disasters. She ordered preparedness pamphlets from government websites.

'Is it bad?' his wife said. She folded the towel and put it on the counter. She patted it.

'It's more of a pinch.'

'Let me see,' Dolly said. She came around the table and looked at his foot. 'It's bleeding a little. If there's blood it isn't a pinch. It broke the skin.'

His wife squeezed the bridge of her nose with her thumbs.

'I've got to get ready,' she said. 'I'm showing a house in a half hour.'

'Go on,' he told her. 'I can handle this.'

She walked out of the kitchen, testing her hair with her fingers.

He took a waffle out of the freezer and slotted it into the toaster.

'Aren't you going to call?' Dolly asked.

'I'm going to eat this waffle first,' he said.

When his wife came back into the kitchen he was eating.

'There's coffee,' she said. She took some earrings out of the junk drawer. She kept her favourite pairs in there, for some reason.

'It's OK,' he said. 'I've got juice.'

'I'm going now,' she said.

'You should go around the front,' he said.

'Why?'

She became so easily annoyed.

'It's still out there,' he said. 'In the yard.'

'What?'

'The squirrel. He was sitting near the fence.'

Dolly rushed to the glass doors.

'God,' she said. 'I think there's something wrong with it.'

'It looked OK to me,' he said.

'You'd better call someone,' said his wife. Then she left, using the front door. It hadn't been opened in a while. It made a crackling sound. It had warped in its frame and needed sanding.

'Do you need the number?' said Dolly.

'I'll call information,' he said.

'It bit you,' said the man from animal control, reviewing a fact. The man had a sweater on over his coveralls. It was the kind of colourful sweater that men of a certain age receive as gifts from their families.

'Right here,' he told the man, pointing down at his foot.

The man didn't look. His concerns were animals.

'Was it doing anything strange before the attack?'

He thought that was a strange word for the man to use, but he supposed it was the right word.

'It was just in the grass,' he told the man.

'And afterwards?'

'It ran away, but only to the fence. See it over there?'

'I see it,' said the man. 'Can you remain in the domicile while I fetch a few things from the van?'

The man came back with a long net.

'Where's the other thing?' he asked the man.

The man looked at him quizzically.

'You said you were going to get a few things from the van.'

'I meant one thing,' the man said.

Then the man went out through the glass doors and put the net over the squirrel. He flipped the net and bounced it to make sure the squirrel went into the bottom. It seemed easy.

It was Saturday, so he'd be alone with Dolly all day. He asked her what she wanted to do.

'Don't you want to wait for the results?' she said.

'I doubt they'll be in today,' he said. 'The animal control guy said he'd never heard of a squirrel getting rabies.'

'They'll have to cut its brain out to be sure,' said Dolly. 'And a tetanus shot. You'll need one of those.'

'I had one last year,' he said. 'The thing with the pruning shears.'

'Oh, right,' Dolly said. She turned back to the TV.

'We're leaving this house today,' he told her.

'OK, OK,' she said. 'I'd like to go to the cupcake shop and I'd like to drive down and see the whale.'

Her mother wouldn't like the cupcake idea. Dolly was getting doughy around the middle. 'She can be weird or she can be chubby,' his wife had said more than once. 'But she can't be weird and chubby.'

He told Dolly OK and to get ready.

'I am ready,' she said.

A small whale had been attempting to beach itself near Sandy Hook for the last three days. It was on the news. When the whale got close to shore, groups of volunteers would coax it back out with their hands and gentle chanting.

On the half-hour drive he stole glances at his daughter. She had chocolate frosting in the corner of her mouth. A few years ago he would have reached over and wiped at her lip with the pad of his thumb, but she was old enough now to keep her own face clean. Thirteen was a hard age, he understood, though it hadn't been particularly difficult for him. The drudgery of school, yes, but mostly he remembered the bright summers, the smell of chlorine, the snap of a ball in a glove. It was easier, maybe, for boys. If you had an interest in sports and could throw and catch, it wasn't hard to get by.

'Do you know we're in the middle of a major age of extinction?' Dolly said. 'Bigger than when the dinosaurs died, even.'

'That's interesting,' he said.

He meant it. He himself often thought along those lines. For example, was evolution still happening to human beings, and if so what would they look like in a million years? He recognised that, at the rate things were going, it might be a moot question. It wasn't something he was going to discuss with Dolly, who worried enough as it was.

'People always tell you something bad is interesting when they don't intend to do anything about it,' she said.

'I don't think there's much I could do,' he told her.

If only his daughter's disappointment was bred by smaller things. He wanted to make her happy.

The wind whipped off the ocean and tugged at their jackets. He pulled his collar up. A gust lifted the hair off his forehead. He felt vaguely heroic, approaching the beach under a sky of steel. Like someone out of a World War Two novel. The sand was coarse and grey, littered with twisted pieces of black wood. It was winter sand. But winter was still a few months off.

A few huddled groups were scattered on the beach. Primarily they wore parkas but a few were in full-body wetsuits. There was only one person in the water, a surfer straddling his board about a hundred yards from shore.

A young woman jogged toward them purposefully. Her jeans were soaked up to her thighs. She had a red eager face. She squinted as if she were visualising her ideals.

'He came in, but now he's back out,' the young woman said. She shouted to be heard over the wind. 'This is my third day.'

'Do you sleep here?' Dolly asked. His daughter was hugging herself against the cold.

'Sure,' said the young woman. 'We take turns combing the beach with flashlights. Twice last night we had to push him back from shore. Nobody's helping. The Coast Guard was here, and then a guy from the aquarium, but they gave up. We're the only ones who believe he doesn't really want to die.'

The flush was dissolving from the young woman's cheeks, a blossoming played backwards. He thought the worst part of being grown was realising how inevitable everything is. He thought how astonishing it would be if he grabbed the young woman and kissed her. He pictured his daughter's astonishment.

The surfer came up from the beach, carrying his board. He was bearded and thin, with wet ropes of silver hair down his back. He stuck his surfboard into the sand near them. On the bottom of the board was written, 'This machine kills fascists.' The surfer put his arm around the young woman, who shivered into his embrace.

'He's out there,' said the surfer, pointing at where he had been. 'He's right out there, waiting.'

They peered at all the grey water. Something broke the surface. It looked like an inflated plastic garbage bag bobbing up. There was a spray, like a sneeze, and then it went under. Along the beach other groups were pointing and calling out.

'You've seen a whale,' he said to his daughter.

You could divide your life this way, he thought. I have not seen. And now I have. This must be what birdwatching was all about. Or maybe everything.

'He's not really a whale anymore,' said the surfer. 'He's been touched by man. A whale's one of those things not meant to be touched by human hands.'

'He's like a dog, then?' said Dolly.

'Something sadder than that,' said the surfer.

'Michael teaches economics,' the young woman said.

The whale bobbed up and blew again.

By the time they got back it was dark. His wife had come home while they were gone, but left again. There was a note on the kitchen table that she was meeting with her memoir group and wouldn't be back until late. She'd drawn a sad face, which meant she hadn't made a sale. Along with the note, she'd left takeout menus from Angelo's and the Chinese place. Dolly said she wasn't hungry and went to her room. She kept several bags of chips in her closet and candy bars in her nightstand. He didn't know how she came by these things. She never asked him for a dime.

He heard her settling into her bed, and then the growls and percussive gunfire of the zombie game she played whenever her mother was out. He took a jar of pickles out of the refrigerator and fished inside with two fingers. There was only one left, and it evaded him. He thought of Dolly as a baby, how he'd been the one who could rock her to sleep in his arms with a pattern his wife could never master. One, two, dip, bob. One, two, dip, bob. The deep reward of her eyelids beginning to flutter and relax, like the feathers of a settling bird. He remembered the squirrel and rolled down his sock. The skin around the bite was inflamed and tender. But he felt the same.

At the beach a single tent was luminous against the night. Inside, the young woman and the surfer made blunt love. The whale, within shouting distance but forgotten for the moment, turned seaward.

On the other side of the world, a tiger padded into an encampment. She bore the smell of shit and meat. In her teeth were strands of lung, liver and intestine, offal of her own young. She entered the light cast by the fire. The tiger's muscles slid beneath her stripes like prisoners behind bars. The three men warming themselves, poachers, stood with shock and fear. This was the very beast they'd stalked for days, whose cubs they had found in a hole beneath the roots of a fallen baobab and eviscerated on the spot, taking what they needed and leaving a small pile of wet guts where each had slept. Now the tiger had come to rip them down and tear out their throats. They found themselves unable to run. Their rifles, leaning against a tree fifteen feet from where they stood, were like distant things in a dream. The fire popped loudly and they heard the heavy shuffling of their elephants in the woods.

And then the tiger, like a house cat, rolled onto her back and offered her heart to her hunters.

SIX

We will celebrate writing and art which is grounded in a sense of place and of time. Our literature has been dominated for too long by those who inhabit the cosmopolitan citadels.

THOMAS KEYES — Roe Deer in May Birch — *Roe deer parchment, birch smoke, birch tar smoke* — *Issue 5, Spring 2014*

This piece has been produced on roe deer skin parchment impregnated with birch smoke using a technique developed by the artist. The parchment is created using the traditional method of curing a skin in lime to remove hair and mucus membranes before drying under tension on a frame to create a smooth, strong, translucent surface. The image is built up in layers of birch wood and tar smoke which is dry distilled from the bark. The entire process takes place in the birch woods of the Black Isle in the Scottish Highlands among the roe deer.

Dark Matter

IAN HILL

Issue 3, Summer 2012

Here is a beginning of sorts; a mossy basin of sphagnum and sundew, juncus and bog myrtle. Here is one of the many places where the river rises, as subtle as the cast of light on the fells beyond. Below the earth, layers of moss are settling into a dark peat, in a process longer than our human patience can imagine, the inexorable accumulation of carbon-rich humus.

From the lip of the corrie, a mountain stream seeps from the bog, trickling over stones dappled grey like the skin of a seal. Gathering pace, the stream hushes through gorges and waterfalls, echoing in these narrow places like wine poured from a bottle. It exposes slabs of grey slate where veins craze the rock like a filigree, twisted bands of quartz, minerals as glassy as the eyes of the wandering sheep. In places, the stream descends into black holes scooped from the stream bed, recesses overhung by ferns and woodrush, seeping with a dampness that persists through the short Cumbrian summer.

It is in these velvety hollows that local farmers found a soft, black mineral below the bole of a tree which had come down in a gale. It left a leaden sheen on their fingertips, a dull iridescence like a beetle's wing-case.

This strange stuff, this *Wad* as it was known, had its uses; it cured colic and indigestion, it could be smeared on the fleeces of sheep as a marking, an identification for the few who strayed from this valley into which they had been heafed as lambs, the fell slopes as familiar to them as the bleat of their own mothers. It sucked in light, leaving a dull burnished glow beneath the flat northern skies. Its similarity to lead gave it another name: *Plumbago*. Neither name belied the fact that it is pure crystallised carbon.

Carbon has two crystalline forms; only an accident of alchemy determines whether the carbon atoms lie in ornate polyhedral structures or flat layers; whether it is clear or grey, hard or soft; whether the mineral is the valuable diamond or its errant sibling, graphite. Both forms have been guarded, fought over, prized. Both are tainted with blood; both have witnessed their fair share of death.

In the north of England, Plumbago became the sought-after substance, a smooth alchemy of slipperiness. It was thought too precious for shepherds, too lucrative to be left in this remote valley. Guard houses were built at the mine entrance, where crates of the dark earthen mineral were packed into carriages and driven along the rough valley road for the ports of southern England; busy wharves where cannonballs were stacked in neat geometric piles, bound for the sea war against France, against Spain, against the nations to which we were bound by familial hatreds. I picture those carriages creaking along the twisting track in the gathering dusk; on the jump seat a bored excise man with his musket aslant over his knees, staring glumly at the retreating mountains. Ink-black against the sky are the yew trees which line this fell side; trees as ancient as war, as old as longbows and spears, as mute as statues. Yews are always blacker than the night against which they are seen.

In a darkened warehouse on a wharf side in Portsmouth or Chatham, Borrowdale plumbago was smeared around the edges of cannonball moulds to give a slippery smooth surface ready to receive the molten iron; a mutable substance, part fire, part liquid, a colour so bright it seems almost an absence of colour, a burning-away of hue and tone into this red-white intensity. Plumbago has one of the highest melting points of all naturally occurring substances. The newly-minted cannonballs slipped from their moulds like a stone from a peach. The smooth surface guaranteed that they would fly further and faster than those of the enemy; such is the futility of the arms race.

By the nineteenth century, cannonballs were losing effectiveness. Swords were recast into ploughshares. The plumbago had a new use, a new name: *graphite*, a name for writing. A newly-literate population demanded the tools to draw, to scribe, to make notes and record ideas. They needed something more convenient and portable than slate tablets and lead sticks.

Carbon and mark-making have long been indivisible; its dark leavings have been known from hearths and tallow candles; it is the black residue that remains when all else is burnt. Neolithic humans smeared soot from their fires on the walls of their caves. Quills cut from the feathers of goose wings were dipped in inks made from lampblack and oil; earlier, in an age when all we had came from the woods and went back to the woods, ink had been made from oak galls ground into a fine powder and mixed with rain water; carbon in its sepia hue. Carbon became the means by which we could express our inner, carbon-based souls. Now, this seam of pure crystallised carbon high in a Lakeland valley birthed a new industry, as neatly as a stone slipped from a peach.

∽

Downriver from Borrowdale, a new mill in Keswick met the growing demand for pencils. The graphite came from the Borrowdale mines, the wood was seasoned Florida Cedar, cut from used railway sleepers. Pencils became the readily available means to write; convenient, less messy, sized to slip into a shirt pocket, for it was men who demanded this need to record, to think in columns of numbers. Demand exceeded supply; the need to increase output forced the industry to reduce waste, to find new processes to save graphite. In 1838, a young entrepreneur and inventor, Henry Bessemer, patented a technique for compressing waste graphite into solid form, enabling pencils to be made from the waste products of the industry. For Bessemer, it was the start of a long career as an inventor; ideas buzzed around his heavy-jowled head. The industrial revolution, for men with a quick mind and access to private capital, was there to be mined for opportunities and wealth.

In the 1850s, Bessemer had moved on from Graphite. He was working on the problem of producing higher-quality steel at industrial scale. The weaknesses of pig-iron and cast iron, the simply smelted metals which had been used for cannonballs in the sixteenth century, were limiting the growth of industry; railway bridges were failing, engineering was hampered by the brittle nature of iron. Bessemer's new process burnt the impurities out of molten iron ore by forcing high-pressure air through the smelter. For iron to transform into steel, it needs an increased concentration of carbon. The alchemy of the blast furnace, the chemical wedding of Iron and Carbon, is the perfect metaphor for the industrial revolution, the ideal union of substances with which we were already familiar, which were available in abundance, and which could be turned to new means of production.

By an accident of geography, West Cumberland had the perfect conjunction of conditions; coal, iron ore, access to the sea. In the same year that Henry Bessemer perfected his Bessemer Converter, the Workington Haematite Iron Ore Company was established. Workington acquired the first commercially viable Bessemer Converter in the world. And it prospered. The steel works grew. In the first half of the twentieth century, it was said that the skies above Workington glowed red all night, illuminated by the unearthly glare of molten steel in the blast furnaces. Strips of molten metal to be pressed into railway tracks thundered on rollers through the cavernous, echoing sheds. Men stripped to the waist, their skin slicked with sweat, shouted above the noise of the furnaces. To enter the steel works as a stranger was to arrive in an imagined hell of heat, noise and light. Here was the industrial revolution in all its soot-handed, swarf-scarred raucous reality. Here was carbon in its pure form, steaming from smokestacks and chimneys, ponding in the limpid air above factories and mills, oxidising silently in our skies.

The steel works are now derelict; an empty space of roughly turned earth and shards of concrete. Towards the sea, gulls wheel over the bank of spoil, the debris of dozens of years of steel making, the earth toxic with the sulphurous residue of chemicals. The world's steel industry has gone elsewhere; to India, to China, to countries whose industrial growth spread across the land on railway tracks made in Workington.

When we die, the carbon from which our bodies are largely composed, from which all life is structured, begins to decay. Carbon-14, the unstable twin, the one which exists only in the fragility of life like an expression of our quickening existence, slowly loses an electron. It mutates into the stable nitrogen-14. I picture these lost particles drifting into the skies like the gentle release of our souls, fading slowly like the light from a dying star. Higher, in the thin and frozen layers of the atmosphere, they surround us invisibly, mingling with the other carbon we have unwittingly poured into the atmosphere; the black residue that remains when all else has been burnt.

Outside the window of my study, I can hear the dense winter rain. It falls heavily on Workington, on Borrowdale, filling the upland bogs, seeping into the rivers until their banks swell with the weight of it. I listen to it running down the windows, drowning the sound of my words scribbled on the page, the flakes of graphite smeared from my pencil as I write.

October Black Isle Pheasant Stew

THOMAS KEYES

Issue 2, Summer 2011

Ingredients

One pheasant
Hazelnuts
Hogweed stem
Burdock root
Ground elder
Wild chervil
Chanterelles
Brown birch bolete *(Orange would be better, but none to be found on the day.)*
Lycoperdon pyriforme *(A woodland puffball I can find no simple name for; Latin has its uses for classification but in terms of description and association the common or folk names of wild foods are more useful and accessible.)*
Potatoes
Onions
Garlic
Cabbage
Runner beans
Kale
Cooking apples

The first thing is to do an autopsy of the pheasant; the more recent the kill, the more damage it's reasonable to put up with.

This one was still warm when I found it, so although the guts had been a little mangled – hit from behind, rather than a nice clean knock on the head – it was still fresh. The meat was good and this recipe involves boiling the hell out of it anyway.

It's easier to skin the bird, rather than pluck it, both practically and to disassemble the characteristics of life more quickly. Remove the head and wings with the skin, then cut all the good meat and fat from the carcass. The meat is the real prize here, along with the hazelnuts; enough energy to compensate for the effort with plenty to spare. The life that flowed through this meal will soon be the energy running me, interpreting the process of life as a component, not a consumer or viewer. The chain is so direct, it can be seen. No wonder they used to worship the sun.

Next, shell and crush a few handfuls of hazelnuts and put them in a pan with some fat from the pheasant. Warm them on a low heat and then add the onion, chopped finely; there's enough oil in the nuts and fat to fry the onion. Now is the perfect time to pick hazelnuts. It is the most human of activities, the one primates are made for, a timeless experience no tool can interrupt. By Poyntzfield, there are loads in the burn. It's clear under the hazel, except for a few ground ivy and dead nettle blow-ins struggling. They hold back the bracken and brambles, creating a series of interconnected glades running up the banks of the burn; deep spaces with one rising side and a sense of intangible length; water flow subtly adding to this impression. We were there before the fall, getting the first choice of nuts still on the trees, better adapted than any other creature around here to take this harvest. No grey disruption; the mice have to wait.

Last year we shared. It was an exceptional year and they fell early and in synchronicity, such a dry summer that the trees by the bottom of the burn, usually damper than hazel prefer, took advantage and produced ten times what there is today. The mice are more frugal this year; it will be different. We picked for two hours. While the trees still clutch them, we have the advantage, but this will all change next week.

I should have painted here last year, when surplus bought the time and they fell early in good light. It will be too close this time; the light will be gone before I have time to spare. Twice now I've promised myself this, and each time the imagined process of the work is refined. Next year, it will be even better. I started with a woodcut, but this year that model no longer suffices, it's a paint job, a drawn-out event that needs its light more than its shapes. Too simple a process to be reduced to a lithograph, it needs raw painterly digestion. As hand and eye pick, they paint.

When the onion looks done, chuck in the pheasant carcass along with the burdock root, hogweed stems, ground elder stems, wild chervil and garlic. This is just the time to dig burdock. A bit late for the rest, though places which have been strimmed or mown tend to have a second flush, hogweed sometimes even flowers this late. Ground elder and hogweed tend to get a bad press. The first, introduced by the Romans as animal fodder – an invasive species in all senses of the word – is now a gardener's nightmare. I've earned a few days work unthreading its endless root network. Hogweed is phototoxic, so will give you a rash in sunlight, but like nettles, once you get around that it's a great staple.

Alter the quantities to taste, but the aim is to get a really earthy, gamey stock. Add enough boiling water to cover and simmer.

While waiting for the stock, take more crushed hazelnuts. Fry an onion with them as before and add in the pheasant meat, chopped into lumps, fat left on. When the stock is ready, add the meat to the pot, along with the potatoes, mushrooms, cabbage and kale.

There's no more enjoyable harvest than potatoes. The schools still have tattie holidays here. There is a real suspense around digging them, since the crop is hidden right until the end. We can only imagine the thrill or terror this crop inspired when life depended on it.

Simmer until the potatoes are softening. Add in two diced cooking apples and the runner beans; simmer until they are soft and you're done.

Eating, this is good food, rich, heavy and thick. Beyond that, it has allowed me to tie a series of fortuitous events and harvests into a meal. The apples worked. Too many chanterelles. A few small stones, so the gizzard must have been ruptured. I didn't pick this up during the initial post mortem, but it tells us more about our bird's final moments. Struck from behind, across the back, suggesting an attempt at flight, or a car with very low profile bodywork – after this, it seems she was thrown into the air, before landing head first on the verge. The relatively closed wings, the completely uncontrolled landing say the death happened at the first impact. Hardly dignified, but quick at least.

It's not a moral choice, eating like this. Of course, everything about the recipe is self-reliant, anti-consumerist, anti-globalisation and the rest, but I don't always eat this well. Every now and then, the kids pester me into a Burger King; XL Bacon Double Cheese evaporating any resistance.

This time, the decision was made for me. I live nine miles from the nearest shop, my car died last week, it's Sunday and the heating oil is running out. My day would consist of beans on toast in three jumpers, if the land didn't provide.

I've just chosen the ability to choose. And when that's the way things go, it's a satisfying experience. It would be possible to go further: instead of mains water and the electric oven, boil rainwater on an open fire. Here we are, surrounded by wood on the estate and forestry.

Free men used to be granted 'estover', the right to collect dead wood. Maybe one in ten of us was free, when this was law. There's always a catch. We're all free now, of course, and can't touch it.

Genuine California Almonds

SARAH REA

Issue 6, Autumn 2015

We waited all winter for the snow that never came. All the mouths gaping.

Where I come from we have always supplicated for water, and now it has begun to shrink from us. This is hurtful because lakes abound and rivers run wild, released from the snow in the high quiet places where it sleeps in the pines. But there are great cities below, and they are dying of thirst. They demand we cement the walls of our granite valleys and hold the water back for them, for without it they will surely die. They flush their toilets with it.

The ground is sinking in the valley. We are siphoning the aquifers too quickly, and they have begun to compress. The great fruit basket needs the water. To feed the almond trees. For the world has become accustomed to almonds. Where I come from the curled peeling bark of manzanita scrub spreads its legs for the lick of wildfire. This last August it burned, burned, burned the paws of bears and they had to be killed because they were screaming from the pain. It burned our swimming holes and now the waters flow soapy with ash and spent retardant. This spring I did look for morels and found instead broken pine cones populated by green siblings elbowing for the newly-found light. I wanted to find the morels. I wanted the destruction to amount to something magical, fed by moonlight. I wanted to fry them with wild turkey eggs and feel, just this once, that we could thrive in abundance by just looking. I will continue to look. I can feel them below the surface. They blossom from flame and detritus.

My knee is welted and weeping from where the oily oak kissed it and where I scratched it in my half-sleep, spreading the poison everywhere. It would have remained dormant if it hadn't felt so good to draw my nails across the blistering flesh. I wish I could slice the welts into crosses and let the burning bleed away. Where I looked for the morels was at the top of the Priest Grade, on the highway to Yosemite. There is a farm being started. There live the wild turkeys and the goats and the sheep and the ticks that the dog brings back into my sleeping bag. There live the poison oak bushes and the dirty beautiful ones who don't do their dishes, they just make sure to keep their plates straight.

They live in trucks while their sheep have houses. They burn rotten wood and dead squirrels the cat brings back. They are building a place to belong.

At the squirrel's funeral pyre I drank too much scotch and said to a boy isn't this just a wonderful place to have a farm? It is at the top of all these roads which could be destroyed so easily. He laughed. I hadn't meant to be funny. With the roads gone all the goats will be safe from the strangers. The pipes which bring water to the fine espresso machines of San Francisco plunge down the hill just two miles away. There are neighbours here with guns. It is indeed a good place to have a farm.

I am anxious that I do not have such a farm. There was a man once, he could have built one for me. He knew how to run the chainsaw. He knew how to change the oil. He knew how to shoot the quail. He knew which leaves were which and the path of Orion. But he never did stare into my eyes and tell me of his dreams, and so I left, because I am a dreamer and I am wistful and I need words, so many useless words, to feel alive.

Then I think, what good will words do me if there are no goats for the milking? I thought, perhaps, if I had the green-eyed children and the cast-iron skillets then I could live without the whispers and the grasping hands and the pounding hearts. But something pulled at me from the notches in my spine and I created scabs to nurse on my scalp and then I was gone. There are ravenous things below the surface. The man could not afford to lose any fingers or toes down there with me. Now it is the dog and I – he watches birds. He is the culmination of centuries of evolution, and he will find me the quail and I will shoot them myself. I do not have the farm but I can fire the gun. In this, I do not feel so weak-hearted.

The ground is sinking in the valley. But we are accustomed to almonds now. They grow in what used to be a floodplain, where the rivers once split their bellies full of digested igneous rock. We fixed it, though. Now the rows are orderly and the great-grandchildren of Rose of Sharon smoke meth in their trailers and hurl epithets at the foreign-tongued labourers who rendered them impotent. I think of Steinbeck, and all the spilled dreams of California weeping into the Pacific. I see coming tides of seagulls and star thistle. In a world so accustomed to almonds, it will grieve us when only the strong remain. I never want to forget the taste of them, hot with oil and salt, shattering from the gentle pressure of my strong, white teeth.

We waited all winter for the snow that never came. Today it falls on the daffodils.

Visitors Book

GREGORY NORMINTON

Issue 3, Summer 2012

we had a wonderful time I loved the trees and the wood ants alek
saw a red squirral but I was to slow please bring back the beaver
and the links they are part of the balance of Nature

> Katy

A most memorable visit! The Caledonian forest is a marvel and
Mr Muir a most instructive guide! Highly recommended!

> Robert and Wilma Dalrymple

I never knew it was so rich: birches and willow and alders and
Scots pine. How lovely, despite the midges! Shocking to think only
1 per cent of it survives. I will certainly come back to the Caledonian forest – maybe see a black grouse leck next time, who knows?

> Margaret from Govan

A beautiful place. We should all love trees and destroying them
so we can wipe our bums is just insane?! Alec, you really made
me think. Next time I will take the train instead of flying, honest.

> All best wishes,
> Sandy Parnell

We have meant to come on this expedition ever since the children were old enough. Glad we made it at last. Unfortunately
the guide, though courteous and informative, rather depressed
everyone with his talk about the environment. We came here to
recharge our batteries, not to feel like everything is doomed. Of
course one can imagine the loneliness of living and working in
this cottage, so far from the comforts of civilisation. Perhaps a
holiday is in order?

> Andrew and Mandy Harrison, Surrey

Cheer up, Alec! It may never happen!

Sunderland Tony

A wonderful location but not as warm a welcome, to be honest, as we'd hoped. The guide (or is he a hermit who keeps getting interrupted?) left us feeling ashamed of our ignorance about woodland ecology rather than illuminated. This was not what the glowing endorsements on your website led us to expect!

Ben & Celina (Ayr)

Interesting place, nice countryside, sorry we have to leave so early.

Vic and Jan Morgan

The forest is of course beautiful though the walk up was rather hard going and some steps are sorely needed. But the bothy where we stayed was dirty, unheated and full of leaves. The food was seriously below par and I did not appreciate Mr Muir's comments about our Land Rover. We did not pay to be badmouthed like this and blamed for all the ills in the world.

Mrs J. McGrath

Alec, you have a beautiful soul but you mustn't take responsibility for the whole world on your shoulders! Remember that the world is a CIRCLE and NEVER ENDING and that you are only a PART of it.

River, Santa Barbara

I am not staying here another minute to put up with this leftwing claptrap. I will be writing an official letter of complaint.

N. Slater

I am simply appalled to learn about the forthcoming closure of this bothy and the invaluable service rendered by Mr Alec Muir. Why is he not appreciated I would like to know? Perhaps because he speaks the truth about the terrible things we are doing to our planet and no wonder when he sees the seasons change out of all recognition. I have come here on holiday every year for five years and was shocked to find my host so brought down by the shortsightedness of his employers. Don't you realise what a jewel you have in your crown?

Norman Stone

*

Never will we forget the beauty of the Highlands in the fall. My family comes from these parts and it was wonderful to walk where my ancestors walked and to see the beautiful forest with its mosses and the lichens on all the trees. My husband is a Rockies man and he said it reminded him of home only it is a lot wetter! Thank you, Highland Fling, for arranging a truly unforgettable weekend. As for Alison, she is a witty, charming and enthusiastic guide: in short, irreplaceable.

Annie Chisholm
(from Santa Fe, New Mexico)

Crawling Home

ROBERT LEAVER

Issue 7, Spring 2015

Dawn begins long before sunrise. The first light comes into the sky at around 4:30am and I'm awake without an alarm at 4:25. My limbs are sore and I've been tossing and turning for a few hours of restless sleep.

I quietly put the suit and boots on and pack my pads into the daypack for the last time. My dog looks at me, confused by this much movement at an hour that is usually so still. I promise him I will walk him when I get back.

Outside the birds are singing in the half dark. A stray person here and there hurries along on their way to a job, or maybe home to bed. Thick cloud cover and cool muggy air.

Over on the next block a whoop and some animated laughter probably coming from people who have yet to sleep.

I walk down Broadway to 149th Street and I sit alone on a bench in the median centre strip facing north, waiting for my wingmen. I wonder if wingmen are needed this morning, but it doesn't matter. My friends are part of this now, and I could not deprive them of a 4:30am wakeup call and a trip uptown to see me home. This morning it is all hands on deck.

Everything feels freighted with finality. This is it. The last time.

Larry, Teddy and Jack arrive looking fuzzy with sleep. I am surprised nobody balked at such an early start. I am restless to begin and I pace as they get ready. After a moment alone I begin my nineteenth and final crawl. I move along on all fours and remind myself to make special note of all the sensations, to store away everything that my senses can hold. This affair is about to end.

At 151st Street a shopkeeper and another man stand and watch me from the door of a bodega. One man asks me where I'm going.

'I'm going home.'

'We all got to go home sometime,' he says.

As I crawl I mull over the spirit of this experience and I recall the origins. This whole thing started a couple years ago with a picture in my head of myself crawling, a laugh, and an overall feeling of desperation and helplessness. I wanted to crawl in protest of the destruction of our planet. I was outraged over the BP oil spill and various other man made horrors. I was going to

During 2014, writer and performance artist Robert Leaver crawled up the length of Manhattan. Wearing his father's pinstripe suit from the '60s, he started at the southernmost tip of Broadway and ended at his home in Washington Heights. *Photographs by Larry Fessenden*

crawl to the White House from NYC with a picture of Earth on my back and a GoPro camera strapped to my head. I abandoned all that for many reasons, most of them logistical. But I could not shake the picture of myself crawling.

All the way up Broadway people have asked me what I am protesting. The impulse to protest, or to crawl, was brought on by a need to do something, to find a gesture, to take some kind of action that felt true to myself. I wanted to find a public act that symbolised our/my collective desperation and perseverance. As far as I can tell the way we are living is driving our world to its knees. With all our tools for making things easier and for finding answers we as a species seem to be more lost than ever. At least that's how it often looks to me.

I imagine the people I've seen on Broadway and maybe people the world over, feeling a weight on their backs, in their hearts and souls. Maybe this weight is the burden of modern life, the burden of being conscious in a world gone mad. Crawling seemed like a way to maybe show compassion or solidarity, to make a living metaphor of this collective burden we all share. Instead of crawling I could have curled up in the foetal position in perfectly chosen locations. But this crawl was never about surrendering. I went down and kept moving, kept pressing on as so many humans are doing every day. The idea has always been to keep on, to get through this journey, to make it home safe and sound. As far as creating a coherent and specific form of protest, the crawl never became that, at least not as far as I can tell.

On a personal level I wanted to stir myself up, to shake things loose and shock myself. Ultimately the whole thing may boil down to an overpowering desire to blow my own mind without blowing up my life. In that I believe I've succeeded.

Anyway, forget all that analysis for now. I need to get present. I am here inside the final crawl and all else can wait.

I am here with my sweat and the gray muted dawn. I crawl strong past Trinity church and the last active graveyard in the city. The graveyard here is on both sides of Broadway. This land was once Audubon's farm. The reason the graveyard is advertised as active is because there is still space available. Call now if you want to be buried in the ground on a hill overlooking the Hudson with mighty oaks moving in the breeze above you and a hawk holding on at the top of a white pine. Squirrels are busy on the ground. Crows are causing trouble and planning something.

I am here now alongside this burial ground, my leather clad hands sliding across great smooth slabs of blue stone.

Off to my left we have the Academy of Arts and Letters, an honour society of architects, composers, writers and artists. They say the honour of being elected is considered the highest formal recognition of artistic merit in the United States. I crawl past that hallowed hall, sweat dripping off my face, drops falling and splashing onto the red bricks.

Now I stop and rise up onto my knees. I can see my building. There it is, just across the street. I've come home so many times, but never like this. I'm glad it is only 6am – less chance of running into neighbours. Not that I care anymore. Not that I ever did. I am definitely not ashamed and not quite proud, not yet anyway. I am completing the task at hand.

I round the corner and arrive at my building's front door. 800 Riverside Drive. Early morning, before seven, so the entrance is still locked. I dig out keys and unlock the door, still on my knees. I crawl in past the empty doorman booth on a long narrow rug. Crawling on a rug is soft quiet. I move through the courtyard across the concrete where my scraping boots echo loud. The lobby floor is some kind of marble and I slide in near silence like I'm on ice. I pull open the elevator door, crawl inside, and press 8. On my knees I am rising. I am being delivered to a higher place, the 8th floor to be exact. When the elevator dings the door slides open and off I crawl.

I am finally on my knees at my door. This dawn arrival must be quiet. I don't want to wake up my loved ones. My keys jingle as I unlock the door and crawl into my apartment, across the threshold. I stand up and close the door as gently as I can. Everyone is sleeping. Dreaming. This is the start of my life after the crawl. Here on Earth. Wide awake. I am home.

No More Words for Snow

NANCY CAMPBELL

Issue 10, Autumn 2016

> *And if the sun had not erased the tracks upon the ice, they would tell*
> *us of [...] polar bears and the man who had the luck to catch bears.*

> – Obituary for Simon Simonsen, called
> 'Simon Bear Hunter' of Upernavik[1]

'Ilissiverupunga,' Grethe muttered. I'd only recently learnt the word. It meant *'Damn! I've put it away in a safe place and now I can't find it.'* Mornings at Upernavik Museum: an endless round of *kaffe* and conversation as local hunters dropped by to discuss ice conditions. Wishing to make progress in my research into Greenlandic literature, I'd asked Grethe, the museum director, whether she knew of any poetry books. But the bibliographic collections held mainly old photographic records of the settlements, and kayaking manuals.

'Illilli!' Grethe called an hour or so later, *'There you are!'* She emerged from a doorway almost obscured behind a stack of narwhal tusks and proudly presented me with a 1974 hymnbook, its homemade dust-wrapper culled from an offcut of pink wallpaper.

Upernavik is a small, rocky island on the west coast of Greenland. At 72° north, it is well within the Arctic Circle, and the museum claims to be the most northern in the world. The region's coastline is described as 'an open-air museum'. That is to say, people suspect there are interesting artefacts lying, undiscovered, everywhere under the ice. No matter that they cannot be seen. They exist, and the empty museum building awaits their arrival patiently. One of the museum's prize possessions is an old motorboat in which, during the short summer, Grethe visits people in distant coastal settlements who claim to have found an interesting specimen, perhaps a carved flinthead or an unidentified bone. As these visits are often combined with trips to distant family members and rarely seem to result in artefacts being brought back to the museum, the institution evidently fulfils a social function, knitting together isolated communities along the shores of Baffin Bay.

During my stay on Upernavik as writer-in-residence at the museum, I wanted to discover more about the contemporary poets of the Arctic. The local people I met denied any knowledge of such activity. Research doesn't always lead in the direction you expect: instead of books, it was my conversations with the islanders and observation of their interaction with the landscape that gave me a new perspective on the practice, and the endurance, of poetry in Greenland.

Grethe's hymnbook was a perfectly logical offering. In Arctic tradition, elevated verbal expression took the form of songs rather than poems. These songs have been roughly categorised as charms, hunting songs, songs of mood and songs of derision. The 'charms' were used in shamanic rituals to cast spells or cure illnesses, and were closely guarded secrets; they could be used, for example, to stop bleeding, make heavy things light, or call on spirit helpers. The other categories were public, being performed at feasts and flyting matches, accompanied by drumming and dancing. When the Danish explorer Knud Rasmussen began to transcribe the songs, he declared that his 'neat written language and [...] sober orthography [...] couldn't bestow sufficient form or force to the cries of joy or fear of these unlettered people'.[2]

The measure of 'poetic' success was that a song was *worth listening to*, as Tom Lowenstein demonstrates (in his translation of Rasmussen's transcription of a song by Piuvkaq):

> I recognise what I want to put into words,
> but it does not come well-arranged,
> it does not become worth listening to.

Lowenstein describes the intense performance anxiety the poet might suffer: 'Forgetting the words, in a culture without paper, would be like losing the song. No-one would be there to prompt. It would be as if the words no longer existed at all.' This fear of forgetting is resonant, considering the losses faced by Inuit culture today, now that many traditional practices have fallen out of common use.

For a long time the Inuit 'did not know how to store their words in little black marks'.[3] They had no inclination to. Had they felt a need to apply their technical ingenuity to the problem of recording language, the course of bibliographic history might have been altered. As it is, publishing technology was introduced to Greenland by Danish missionaries during the late 19th century. The printing press preserved some legends, but the songs – because of their strong shamanic connections, not to mention occasional explicit content – were suppressed. The drums used in shamanic rituals were burnt in an attempt to oust 'heathen' beliefs, an act as sacrilegious as a book-burning in Europe.

Hushed and drumless, the Danish colonists tried to locate the rich sounds of *Kalaallisut*, the Greenlandic language, within their known orthography. The Roman alphabet was introduced to facilitate printing with conventional metal type imported from Europe. Kalaallisut, the standard dialect, is caught between cultures, one of the few Eskimo-Aleut languages to use an alphabetic rather than syllabic orthography (compare its close relative, Inuktitut or Δᴐᵇ∩Ɔᶜ, found in Nunavut and the Northwest Territories, Canada). Yet what impact has two hundred years of printing made? In 2009 the *Unesco Atlas of the World's Languages in Danger* designated Kalaallisut as being 'vulnerable' and predicted that the North and East Greenlandic dialects will disappear within a century.

When I began to learn Kalaallisut I had to ask my teachers to write the words down. They were bemused that I should find this more useful than hearing them spoken. Each time a word was written it would be spelt differently, and so the bemusement was passed on to me. Grethe told me that schools are not overly concerned about spelling: little children are bamboozled by the long words, and surely it is understandable that they get lost in the middle and miss out a few syllables? Teachers are more inclined to indulge the children than instil superficial spelling conventions.

Many of the islanders found expressing themselves in writing challenging. Speech is still the touchstone for communication: mobile phones and Skype are just as popular in Greenland as they are in the UK, whereas emails are approached with even more dread. It seems inevitable that future Arctic archives will be as sparsely furnished as those of the past.

I began to find English finicky and prim in contrast with Kalaallisut. As though they were knucklebones used in a game of dice, I shook up my tiny words and scattered them before my audience, having little influence on the score. Kalaallisut is more densely woven than English, with its smaller alphabet (18 letters) and polysynthetic words. When it is spoken, the suffixes are uttered so softly that an untrained ear cannot hear them. Sentences seem to trail off into silence.

Kalaallisut will use a single word to express a concept that English tiptoes around with a phrase. I was delighted to find signifiers for the sea rises and

Houses look out over Disko Bay, in Ilulissat, Greenland. Ilulissat means 'icebergs' in Kalaalissut *Photograph by Nancy Campbell*

falls slowly at the foot of the iceberg' (*iimisaarpoq*) and 'the air is clear, so sounds can be heard from afar' (*imingnarpoq*). The language is famous for its many words for snow. This wide vocabulary for environmental conditions is of fundamental importance in understanding the Arctic ecology. As Barry Lopez points out in his book *Arctic Dreams*, contemporary scientists who arrive in the Arctic to assess climate change without a grasp of Kalaallisut risk being as crude as the early explorers who rushed to make their conquests of the North Pole without using established Inuit techniques for transportation and survival on the ice.

A map by the cartographer R.T. Gould in the National Maritime Museum in London delineates the last known steps of one such expedition led by Sir John Franklin, an ambitious Victorian quest to find the North-West Passage (1845–8). Gould's map depicts a land marked not by geographical features but by ominous 'x's: caches of letters, pemmican and bones found by search parties. These clues to Franklin's disappearance, linked by a red dotted line, eventually peter out in a question mark surrounded by blank paper.

The North-West Passage can be located by satellite these days, and few uncharted regions remain for those wishing to make their reputation as explorers. Yet despite advances in knowledge, Arctic geography still challenges the complacency of the modern traveller. Much of the visible environment is

A hunter's illukasik, high on the rocks overlooking Disko Bay, Greenland *Photograph by Nancy Campbell*

characterised by transience. The Pole is a shifting entity rather than a fixed point. Icebergs drift along the horizon, an ever-changing mountain range. The shore-fast ice forms an increasingly unpredictable border between land and sea; it disappears almost as fast as the tracks that pass across it. The geographer Nicole Gombay writes that these conditions 'require an awareness that the future cannot be predicted. As a result, people must focus on the present. Inuit have often told me, "Today is today, and tomorrow is tomorrow. Don't bring today into tomorrow, and don't bring tomorrow into today."'[4] People's distrust of fixing future plans is balanced by 'an ability to let go of the past'. As Heraclitus might have said, it is impossible to step on the same ice floe twice.

Some mornings when I sat down to write at my desk overlooking the harbour, the sea outside my window seemed like a 'black cauldron covered with dark frost smoke' (as Robert Scott once described the phenomenon in his Antarctic journals). Other days, it was hidden by ice, and I watched the hunters make their way across the perilous expanse until they were just little black marks in the distance. The shadowy figures stepped carefully, pausing often, and tested the ice with their chisels before putting any weight on it. They were adept at interpreting patterns and sounds in the ice, which told them where to step to avoid falling into the freezing water. Each man's understanding of the ice was essential to his survival. (Once upon a time, the intense dangers faced during such expeditions had inspired the composition of songs, and even provided a metaphor for the process of composition: in a common trope, 'the right words' are as elusive to the singer as a seal or a caribou.)

The hunters' ramshackle workstations awaited their return. These *illu-kasik* had no walls, no roofs and no doors. There was nothing to obscure a hunter's view of his terrain, and nowhere to hide a secret. Domestic objects were left to rust under the open sky. The snow was a part of these skeletal structures as well as their backdrop; deep drifts were conscripted as tool racks. Ladders were lashed to the upright timbers but rather than providing a means of ascent they held struts together or secured them to the ground. Green twine wound about the cornices in endless orbits that stood in for more sturdy knots. The whole island appeared to be held together by an armature of twine and chicken wire beneath the snow.

Illukasik evolve. Beams are nailed to the joists, clothes racks tied to the beams. Sealskins are sewn to stretching frames and fish are hung up to dry out of reach of ravenous dogs. An accumulation of clothes pegs, knives and

Kayaker in Ilulissat Icefjord, Greenland *Photograph by Nancy Campbell*

beer bottles adds a distinct signature to each hunter's creation. Between snowfalls, the outer boundaries of the illukasik are pitted with holes cast by phlegm, drops of oil and cigarette butts. Fresh lines of blood are traced across the island nightly as seal carcasses are hauled from the successful hunters' plots to waiting kitchens.

Sometimes, silhouetted in twilight, the illukasik looked like creatures rising from the sea. In these manmade objects I sensed something more than functional architecture. Folk tales describe hunters who created living monsters, *tupilak*, from sticks and stones and breath. The traditional Inuit religion is animist, and the culture is strongly influenced by the belief that an *inue* or soul imbues every material thing, from a rock to a harpoon head, informing its purpose. And so, as the wind howled around the illukasik, I thought of them as expressive marks on the landscape, almost akin to song. While Inuit songs were intensely personal, and singing another's composition without crediting the original author was frowned upon, the singers employed respectful variations on traditional themes. The *ikiaqtagaq* or 'split song' was a conversation over time, its lyrics added to, and developed, by successive singers. Likewise, the design and materials of these improvised buildings diverged little from those I had seen in old photographs in the museum. Here was the continuation of a creative tradition that I sought.

When you store something away in a safe place, there's always the danger you won't find it again. Perhaps it is simpler to accept loss at the outset. The absence of printed language in the Arctic seemed to be as potent as the more tangible literature I had grown up with. I wondered whether a poet writing in English today could be active without publishing, and even whether there might not be a case for silence as a poetic stance in a culture so unremittingly orientated towards self-preservation and self-promotion?

With these thoughts I turned from the museum's bookcase (or, as I had learnt, *illisivit* – the root word of ilissiverupunga) to the gallery vitrines. There I found evidence left by earlier visitors: barometers and log books from explorers' vessels, and the highlight of the collection – the Kingittorsuaq Runestone, engraved with a short text by three Norsemen around 800 years ago and left in a cairn on a nearby island. Only the men's names could be read; the second half of their message is lost, written in mysterious characters that can't be deciphered, even by experts. The truncated story of these Viking travellers is emblematic of the history of the Norse in Greenland. None of these settlers would survive the 15th century, in part because they were unable to withstand the cooling climate of the Little Ice Age.

With the media saturated by images of the Arctic, it seems no longer necessary to convey its appearance, but rather the timbre of its many voices. The anthropologist Edmund Carpenter suggests that in cultures where transience is more evident, process is valued over preservation:

> Art and poetry are verbs, not nouns. Poems are improvised, not memorised; carvings are carved, not saved. The forms of art are familiar to all; examples need not be preserved.

> When spring comes and igloos melt, old habitation sites are littered with waste, including beautifully designed tools and tiny carvings, not deliberately thrown away, but, with even greater indifference, just lost.[5]

It is increasingly apparent that our planet, including all its museums and libraries, is facing a devastation even more extreme than that of the great Alexandrian repository. Gombay addresses Western society as well as that of the Inuit, saying, 'In the face of knowledge that ultimately we are at the mercy of forces over which we have no control, how are we to react? We can choose to ignore such awareness – dig in our heels and do all that we can to find a means of establishing supremacy over the essential instability of existence, or, we can give in to it and accept that our experience is ephemeral.'[6] When the last of the ice has melted, the vanished tracks upon it will be the least of our concerns. No-one will be there to prompt. It will be as if words never existed.

Notes

1. Quoted in Hansen, K. *Nuussuarmiut: Hunting Families on the Big Headland*, Meddelelser om Grønland, vol. 345: Man & Society, vol.35, 2008, p. 146.
2. Rasmussen, *Eskimo Folk Tales*, Kessinger Publishing, 2010.
3. *Ibid.*
4. Gombay, N. '"Today is today and tomorrow is tomorrow": Reflections on Inuit Understanding of Time and Place' in Collignon B. & Therrien M. (eds), *Orality in the 21st century: Inuit discourse and practices. Proceedings of the 15th Inuit Studies Conference*, INALCO, 2009.
5. Carpenter, E. S. *Eskimo Realities*, Holt Rinehart and Winston, 1973, p.57.
6. Gombay, *op.cit.*

SEVEN

We will not lose ourselves in the elaboration of theories or ideologies. Our words will be elemental. We write with dirt under our fingernails.

LIONEL PLAYFORD – From Calvert End to Little Dun Fell – *Peat, clay, ink, wax crayon, wind, sun, rain* – Issue 9. *Spring 2016*

This drawing was one of five I made whilst sitting in the cushion soft moss of a peat bog in the North Pennine hills of northern England one bracing autumn day in 2013. The view is towards the source of the river Tees where water laden westerly winds drop their Atlantic load on a line of hills that run north-south; the so-called backbone of England. Ever since the retreat of the great ice sheet 10 000 or so years ago this climate has nourished both peat bog and forest and now I scoop out a handful of brown paste from the base of a peat hag where it meets the glacial clay base and draw with the partially decomposed remains of that plant history the very thing that created it. The circle is complete and I feel it powerfully in the movement of my hand across the paper.

Prognosis

GLYN HUGHES

Issue 2, Summer 2011

I am told that my atoms assembled into a bomb
exploded in Space would outshine the sun.
But I am thinking of cremation, scattered on water.
The Atlantic wears even granite into atoms
and energy is freed from its prison of substance at last.

When my ashes flow on the stream maybe the salmon
will take a mouthful through the Atlantic,
my threads will wriggle even backwards in time
as happens, says quantum physics

                        and who will know but God,
and perhaps my descendant disturbed one night
by thoughts and dreams that seem like memories?

Osiris

SYLVIA V. LINSTEADT

Issue 9, Spring 2016

How to meet an angel at the desert crossroads

Set up your easel there, by the creosote, on black stones flecked with the white fossils of shells. Set it up facing north, facing the ridge layered red-orange-cream-black, where in your dreams white coyotes stalk and howl. Set up your easel, your heart, on its three spindly legs. Open your cigar box of paints after staring at them, unable to make yourself begin, for five days, only to find them hard and dry as the ground.

Do not give in to despair, to the emptiness without paint, without that smoothing of brush to canvas turning the land around you to a fire in your chest. Walk slowly barefoot toward the place where two washes make a cross-roads. It is not far. Watch for rattlesnakes sleeping arrow-headed under the sagebrush. Stand at that crossroads, where the floods move with the rains. Now the washes are just rivulets and trenches made in stone and sand.

A man will appear, and you will be afraid, but you will have no need to be, not in the way you think – a woman alone in the desert, a man coming toward you. You will see his skin is dark blue as the night and just as covered in speckles, silver-bright as stars. He wears a tall dark hat and rattlesnakes follow at his bare feet. He has two wings like the piñon jays, grey-blue, slightly iridescent, scruffy.

'What would you trade me for a box of paints, sweetheart?' he says, and it is not a hiss like you expect but rich and deep and blue.

Of course you've heard of this sort of business, and aren't sure you want anything to do with it. But he pulls a box made all of coyote bones from behind his back and inside are glass jars of pigment, mixed from pockets of miraculous blue sleeping in the earth, reds and oranges and sage greens too.

'My first born,' you say, as it sounds like the right thing, and you mean painting, not child. He understands, and agrees.

When you have finished it, which takes you all night by candle and part of the morning, a silhouette of that ridge and the sky going dark, luminous and sun-bleached and also somehow the ridge of your own dreams, a spine you touch only briefly, you brew up a cup of coffee over a fire in a ring of stones. Your cup is tin, blue and speckled. You go down to the crossroads and leave the painting.

By night, the coyotes are yipping. They surround the painting. They eat it as they would the carcass of a bighorn lamb, snarling, fighting for the deepest veins of colour, blood on their teeth.

While you sleep, mice gather around your bedroll. You dream inside your blue rustling sleeping bag, on top of the old Egyptian blanket, wine-dark and woven with flowers, that your uncle who lived in Cairo sent you when you were a little girl. You dream of a house in the canyon just beyond the one where you sleep, a house made of old adobe and the bones of animals long extinct – the American camel, the giant sloth. The mice – a cactus mouse, a piñon mouse, a Great Basin pocket mouse, a chisel-toothed kangaroo rat, a brush mouse, a southern grasshopper mouse – harvest the little bits of your dreams from your desert-lank hair like mesquite seeds.

You don't notice, only that you feel bleary-eyed when you wake to the dawn slanting across the eastern ridges, basalt and the memories of volcanoes. It is late October, cold until the sun moves full into the wash. Come daybreak, you build another fire, heat more coffee, use the last of the milk you brought in a jar, measured out for seven days, thinking you'd stay no longer. This is the seventh morning. A handful of trail mix and an apple left. You paint again, what you see. The colours are live as cactus mice and the great flocks of piñon jays, live as sun drying dew off your sleeping bag.

The journey down

At home, before, in the city, you sit at a computer and process donations, write grant proposals. It is not a bad job. At 5.30am you make coffee in a yellow pot on the stove and it smells nutty. Sometimes you add cardamom. You take it out onto the little porch and look for the sunrise, or for a star, or the small beginnings of dove coos. When you come in, you try to paint at the easel you have set up in the laundry room, coffee hugged one-handed to your chest for warmth. You only do it at dusk and dawn, when the sky and the sun are changing, the world is unformed, and a thing like painting is acceptable. Your heart smolders at the edges, keeping you warm through the night, and then the day, that way.

On weekends you take hikes on the nearby coastal mountain. You like the smell of the coast live oaks, the bays. Sometimes on Friday evenings you bring a book and buy a pint at the local pub. Sometimes men try to sit by you, buy you another, and you smile but shake your head, keep reading. You'll know him without even looking up, you assure yourself. And he probably won't be at the pub.

There comes a morning when you rise, the usual hour, grind the beans, set the pot on, and find that you can't do it, not one more day. You find that when you sit to paint, your heart does not smolder at all. It is cold as ash. You don't feel like sitting for one more second. All at once you want to kick the door down. If you have to stare at the bright screen of your office for one more day, you will break it right there, with anything handy – stapler, heel of your shoe.

It feels like an animal rising up in you, furred, clawed, green-eyed. You slam your hands once into the wall, which smarts the soft skin of your palms. You drive to work. You do this for several more months. The swallowing of something clawed every morning, cold at the edges of your heart no matter what you paint. The expensive gauche and new set of oils that you buy for yourself don't help.

There comes a day when you wake up, go to the yellow coffeepot and find that this time you no longer fit inside the square walls of your house at all. The ceiling feels as if it is pressing on your shoulders. The coffeepot makes you want to cry. Your mother gave it to you as a going away present when you left for college ten years ago. You throw it through the window. The glass breaks and shatters like rain. The coffeepot lands in a bush. Shaking, you think about how glass is made when lightning hits sand, and what a miracle, truly, electricity shearing the grey sky is.

You call in sick and go get the yellow coffeepot calmly out of the bush amidst the glass shards. Then you pack your old Volkswagen van, red as dried blood, and drive south for ten hours. It is the 21ˢᵗ of October. The land drains of green and turns gold, then dun, dry and rocky.

You go to the place the sun loves best in all the continent, the lowest bowl of land where that star can lay his body down big and unchecked: Death Valley. You like that once it was neck-deep in salt water and prehistoric fish, spiralled gastropod fossils left behind. You like that the driest and hottest and most barren place once held an ocean in its arms.

You worry that you are having what doctors might call a psychotic break. Well, you decide, that might be, but you don't particularly care. You have always had a mistrust of doctors, and diagnoses, and little colourless pills. As you drive through the desert, waiting until you feel ready to stop, passing wind-worn ridges and moon-pocked canyons, your tyres stirring up dust, you think maybe the invention of the car was the ultimate psychotic break and laugh to yourself.

When you get out of the car the air tastes dry and so clear you think you can taste the blue sky in it. The ridges are barren rock, cragged, the bones of a place in its essence. The valley is miles wide, the mountains bigger by several times than the green forested ones you are used to. It is so big you could fit the whole city inside, and then some. Here there are only flat sandy washes, creosote, a herd of bighorn sheep in the distance. The mountain ridges are the closest thing to time itself that you've ever seen.

You pull on your framepack and leave the car behind, then walk north, carrying several plastic gallons of water in your hands until the weight digs into your palms. You have to make two trips with the water.

How to win the milk of a bighorn ewe

Around noon you go to the crossroads again. You want proof that the blue-skinned man is not part of your possible psychotic break. A few of the scraps of your canvas are scattered in the sand, like bits of flesh, with blood at their edges. You don't know what to make of this and turn away, but you are hoping he'll come back to explain. When you think of that, the feeling in your chest is the same as the first night on the rough dry ground in your blue sleeping bag, lonesome but content, staring up at the bowl of night, so dark and so full of stars the sky itself felt like one great faceted mineral, shimmering. And the Milky Way, for the first time in your life, looked like just that: a river of white, the milk of a million mothering stars.

You turn back to those canvas-scraps and pluck one of your hairs out at the root, curly and dark like your aunt's, a tiny spiralling staircase. It's not much, but you remember a class fieldtrip, age twelve, how the lady leading it said whenever she picked a leaf, a flower, she left a piece of her hair. A fair trade, she said. At the time this concept seemed loony to all of your class-mates, who sniggered. You did too, but on a level beneath the self-doubt of puberty, you thought this idea magical, heady, this assertion that plants might be as valuable, as feeling, as people. You wondered, however, if a person did this all the time, would she go bald?

You place the hair impulsively among the scraps of your canvas. It feels surprisingly good to leave it behind. Tangible, as much a piece of you as the thin sagebrush leaves are to their stems.

On the way back to your three-legged easel, you wander up the narrow canyon from your dream, half expecting to find a house of giant sloth and camel bones. At this rate, you think, why not? Before, you had read about the ancient fossils of forgotten beasts. How they are still lying under the sands and dry stones of Death Valley, which some million years ago was near tropical, thick with vegetation.

Millions of years before that, ocean water smoothed the stones, but now they are sun and wind torn. Your boots are covered with scratches and a tear on the side. You climb over a boulder, keeping your feet to the sandy areas where the flash floods have made paths and the bighorn sheep have followed. The middle of the canyon widens. It feels like an almond-shaped bowl, holding.

You feel a tingle in your chest and look up to see bighorn sheep, a whole herd, hazel and chocolate brown, wool so close to their bodies it has no curl at all. They are sturdy beasts with clever amber eyes. The rams have huge horns lined and whorled as the gastropod fossils. The ewes have shorter horns, no dramatic curve, hardy and sharp. They watch you from a steep path they've made in the far canyon wall. You cannot read their faces. Desperately, all at once, you wish that you could. You lift a hand. They look to their leader, a stout ram with horns that curve far out on either side of his ears. He snuffs the air, stamps a hoof, moves away. They all follow, clip-clopping, sun-burnished. You see a mama with a young one, fur still a gold glow. Her teats sway. The lamb stays right against her legs.

Back at the easel, you paint them into the half-finished canyon on your canvas. The ochres and mustards and browns of the paints in their glass jars in their coyote-bone box seem to be made, under your hands, of sheep wool, of sheep hooves, of sheep horns. You wonder what they eat. Where they sleep. How they speak to each other.

You're out of milk. Actually, you're out of everything except coffee and a small sack of oats. You decide it is wise to leave the next morning, not particularly wanting to starve or, worse, risk dehydration. The plastic gallons of water are empty except for the bottom of one. It is hard not to think about the drive home. Back inside each precise box. Briefly you imagine that the streets and the houses will feel little and insubstantial as tin cans or dolls' toys, the fences and lawns and sidewalks painted on, a stage set hiding the real thing underneath – muddy creek and moles and that big sleeping creature, the bedrock.

You decide you'll leave empty-handed. The bighorn sheep painting will stay behind too, in case he comes back. It seems suitable to take the paints. The paintings are as much his as yours, because of the trade, but the paints, they are your gift, like canopic jars, full not just of pigment but of necessary organs.

By his, you wonder, walking down to that crossroads again, do you mean his, or the desert's?

You leave the painting, sheep-tawny and spiral horned, in the wash. At the last minute you painted one of those horns with star-speckles like the semicircle of the Northern Crown constellation, *Corona Borealis*, you've seen every night at the darkest hour. In the sand, that horn gleams.

Back at your camp you build a final fire. The sun goes down softly. You stare into the flames for a long time before remembering to put the water on to boil, and your dinner with it: oatmeal. Not very appealing. The flames and embers seem to make orange canyons, washes blue as river floods. You wonder if you will be able to hold all of this when you go back, and feel like crying, confused. You thought a week in the desert would be like recharging a dead battery. That simple. Unplug, full bars. Instead you feel pulled apart into many little pieces, scattered about and used by the piñon mice for their nests. For good measure you tug another hair out and throw it in the fire. You sit up, face hot from the flames. From here you can smell the juniper wood smoke, sweet and spiced.

When your eyes adjust to the dark you see a figure at the far edge of the fire, sitting on a sharp stone. He has two grey wings, blue-tarnished as the piñon jay's. In the dark his skin is darker, and full of stars. The firelight on his face and hands turns them indigo. He has kind, human eyes, like a friend sitting down companionably to help make dinner. Around his feet are dozens of desert kangaroo rats, eyes black and glinting, eating at oats stolen from your little sack when you weren't looking. Instead of the black top hat, he is wearing a flat, round hat made of kit fox fur, like the hats you've seen in photos of Siberian nomads.

It's a chilly night. You look at his wings and imagine the cold of clouds. You lift a hand, like you did to the sheep, and smile. He smiles back and holds up a ceramic jar, gestures for you to come take it. Then he pulls a brace of jackrabbits from over his shoulder and skins them right there with a knife sharp and clear as ice. He throws the offal behind him. Coyotes, you can hear them, fight over it. Before you can even speak the jackrabbits are on a spit over the fire. You pour the ceramic jar into the oats. It is bighorn sheep's milk. It smells of grass and sun. You are laughing.

After a moment you say, 'Did you like my sheep painting?' trying to make normal conversation, like one does over a campfire to a friend. He smiles again.

'I didn't even see it before the bighorn sheep got there. Chewed it to pieces like the most delectable of springtime grasses. They lay down around it to let it come up again as cud, and it turned all the ewe's milk thick and sweet. They gave their milk to me for you. I didn't take it.'

You think of the Milky Way, and the ewe with her lamb close to her legs. You come around to his side of the fire with the pot of cooked oatmeal. Both of you eat out of it with two spoons. You are surprised to see him chew and swallow. Yesterday he looked like he was made to eat only cloud and rainstorm and desert wind. Now he is making contented, hungry sounds. When you lean near he smells of nothing except the night, near and crisp, juniper-wood spiced, dark. You eat the jackrabbit meat with your fingers. In this moment you feel that if you could sit here, leaning over a pot of oatmeal by a fire and licking rabbit fat off your fingers, October-dark, ringed in red ridges and dry washes, beside him, always, this point the axis and your whole life spinning around you, not linear but spoked, you would be happy.

How to bring on the first desert rains of autumn

The kangaroo rats leap to your ankles, munching dropped bits of oatmeal. They are sleek, their tails balanced. You start to cry. The tears get in your mouth with the jackrabbit and the oatmeal. You swallow, choking. The tears pool at your chin as you cry more than you ever have in your life, breaths ragged. The salt starts to sting your face. You feel like you are crying out your bones, like your sobs are starting to howl. The man holds out his two dark speckled hands and they widen and widen and deepen like a basin to hold all your tears. They become as big as the whole night.

'I thought it smelled like rain,' is all he murmurs, tenderly, holding your tears. When you are finished and hiccuping, he throws his cupped hands upward. You duck your head but no water falls back down. A cloud passes in front of the moon. You lean against him, only able to breathe through your mouth, little whistles. The kangaroo rats have gathered in your lap and you stroke their heads.

He holds onto you and you lean nearer, find your nose against his bare chest, where the fabric of his shirt is open. There is no hair on his chest, only smooth darkness, a dry sweet smell, speckles of light. Your lips open and press there. You are not this kind of woman, but it seems the only thing in all the world to do, and you don't even need to look up to know it.

By the fire, as it turns to hot and shifting embers that flicker with towers and ancient sandstone cities, camel caravans and ridges sharp as vertebrae, the man lays down your old Egyptian blanket. His bare arms dark and flickering can barely be distinguished from the night.

You make love on the blanket on the hard ground, your spine pressing against the rocks, his spine a thing you are uncertain of. You can feel the bumps under your hands but, eyes closed, he seems as huge as desert mountains and what happens between you is a salt sea, a sky full with the milky teats of stars, his grey wings grazing and grazing at your hips.

Somewhere toward dawn, curled up against him under your blue sleeping bag, you smile thinking of who you were yesterday, and who you are not now.

'Don't be afraid,' he whispers, awake beside you. You are not sure what he means, but you nestle near again, thinking no, no, I am not. How can this be? I am not afraid.

Just as the sun is lifting the eastern hem of the world a shade paler than indigo night, you hear a yipping, a snarling. You know it is coyotes before you open your eyes. You've been dreaming of them again, bone-white, walking the white spines of the mountains, looking down at you. The snarling sounds very close. You open your eyes. There are at least six of them, maybe nine, big and healthy though their desert diet is lean. Almost wolves, save that skinny sharp face, eyes slightly closer together, big cunning ears. Their teeth are bloody.

When you see what they've done, you vomit onto the black sharp ground.

He is ripped into pieces, blue-skinned, red-muscled. They are snarling not at you but at each other and at the pieces. There are forty-two pieces in all – head and hands and fingers and toes and bits of leg and each organ separate, surgical. His heart, you see it at the far edge of the fire, is blue. His blood on the ground is not quite red, but rather a rich brown-black, river silt. His wings are whole, damp with the mouths of coyotes. Not all of him is there. They've carried off important organs, his liver, parts of his calves, his penis, his ears.

You stand up shaking, sick, and yell and throw rocks and the coyotes growl but back away, a tight group of seven, shoulder to shoulder like one big beast. They run, leaving you there alone.

The sun has turned the east ridges pale blue, but it is still somewhere behind them. Cold and naked, you put on whatever you can find – an old wool skirt, a sweater, your boots. This is a different you operating, the one you were not yesterday but are today. This you will not throw up. This you drank bighorn sheep's milk and held the blue-black night and the sharp spine of the mountains against her breasts, winged, his kit fox fur hat pillowing her head.

This is not the same as a crime in the city, a grotesque murder discovered in a garage. This is like gathering mesquite pods to grind and bake into breads round and bright and caramel-rich as suns. You don't know where this thought comes from but it does and you move deliberately, calmly, placing each piece of him into your big red woven basket with its leather handle, where you had been keeping your clothes and the glass jars of paints. Those jars are still in the basket.

The other you is there too, holding a blue index finger, cold, holding a wet kidney, that big blue heart. The other you is heaving and crying, dropping tears across the ground.

You don't know how long it takes to pick up each piece. Your hands are dark with blood and the basket is heavy in your arms. The sun still hasn't risen over the ridges.

You notice that pieces are missing and feel despair. Somehow it is clear to you that you must find every last scrap of him. You sit down on the rough ground and want to cry or scream but then you see a little desert woodrat holding a small and perfect blue pinky-tip in his paws. You lean toward him, whispering, and he scampers off, his furred tail flickering. You follow slowly with that heavy basket. It leaves a dark seeping trail, as if of wet silt, behind you.

The woodrat leads you to that canyon, the one you dreamt of, the one of sheep. The sun does not fill the sky because the sky has become full of dark grey clouds. In the dim light you trip often, scuff your boots, almost drop the basket. His wings are unwieldy and large, balanced just under the handle. You imagine carrying only a bird and not a man. The big slate-blue feathers brush your hands.

In the centre of the canyon, in a sandy clear place surrounded by creo-sote, desert mallow, cholla cactus, more mice and rats are gathered than you can count: desert shrew, Panamint pocket gopher, pygmy pocket gopher, great basin pocket mouse, little pocket mouse, long-tailed pocket mouse, chisel-toothed kangaroo rat, Panamint kangaroo rat, Merriam's kangaroo rat, desert kangaroo rat, western harvest mouse, cactus mouse, deer mouse, canyon mouse, brush mouse, piñon mouse, southern grasshopper mouse, desert woodrat, bushy-tailed woodrat, house mouse. Each subtly different in shape and size and whisker and paw and colour, like tawny pieces of the desert itself, its most essential prey, food for all.

They've made a bed of sagebrush, jackrabbit down, yucca fiber. Cof-fin-shaped. In a neat pile: two ears, liver, calves, penis. The mice are a seething carpet of gold and dun and grey, tiny paws, quick tails. They come to the basket. You start to take him out, bit by bit. They show you how to do it — head by neck by clavicle by bicep by forearm by hand. Stomach, intestine, kidney, liver. You lay him out until he is forty-two pieces whole, jagged as a shattered window. All that's left in the bottom of the basket are the glass jars of pigment. His silt-dark blood has oozed inside the jars. You take each one out, dip your fingers in. Mix your own saliva too, to make it wet enough, and paint along each seam of him, each tear, as if the paint were glue.

The mice and woodrats and kangaroo rats hug their bodies against him, as if to keep him warm. A zebra-tailed lizard watches from a rock. Overhead, black ravens circle and you work faster, painting green and ochre, red and blue. The pieces of him are not speckled silver now, only the darkest indigo. You can't paint the wings on as they are tucked beneath him. Instead, you trace the feather-tips with a yellow pigment like pollen. The final touch.

Rain starts falling then, sudden and big and without warning. The drops begin to rinse the pigment away almost immediately. You yell up at the sky. Thunder shakes and booms. The mice run one by one for cover under stones and creosote and the edge of your basket. You see a bighorn sheep silhouetted on the top of the far canyon ledge, feeding, unconcerned at the rain on her short wool.

The smell that rises from the earth is baked bread and sage and a bodily musk. It is so full in your nostrils and mouth that breathing feels like drinking. You watch the rain rinse the paint pigments down into the lines of his forty-two-piece body, cracked but fit together perfectly, like dry earth waiting for moisture.

You lay down on your back beside him in the rain. Spread your arms and legs wide, like making a snow angel on winter vacations as a child, except here, now, this you, the motion feels like a handshake with the wet body of the ground. You reach your hand out gingerly to cover his, each loose finger. You don't think about what you are doing here. You don't think about psychotic breaks or the journey home.

You are the first desert rain of autumn. You are the roots of the creosote and desert mallow far below, that have spent the summer drinking ground-water fifty feet deep and are now fanning their rootlet hands to catch all this rain. You are the sky grey blue and opening.

Lightning sears down. Thunder. Only a second between. Again. You count. The lightning is close enough that you should be worried but you lay still. Up the wash a half mile the lightning strikes straight down into the sand and forms a crooked tube of glass, shining and dark inside.

It doesn't take long to flood. The rain is heavy. A sound begins, a roar as of large waves. You can barely sit up before the water is rushing toward you, a churning muddy white-capped flood. You roll to grab him, lifting him against you without thinking. The flood is upon you. He is all one piece in your arms as you swallow water and together are thrown upward, tumbling, to the surface.

The floodwaters seem to have a revivifying effect on him, like a bolt of electricity. Suddenly you are not holding, you are held. Great grey wings shake above the surface of the water as you are whipped and tossed, avoiding rocks. He lifts the two of you on those big piñon jay wings. Water drops fall like stars from the feathers. You don't make it far together, just to the ledge where earlier the bighorn sheep perched, feeding. You huddle near, both cold, and watch the flood.

It mesmerises you as it tears the desert canyon into channels of silt. Beside you he is breathing deep, not in alarm or shock at being all of a piece again, but deliciously, a fresh noise, a leaf unfurling in the morning. You look at him, his arms resting on his knees, knees hiked up to his chest, and see that his body is a tracery now of green lines where you rubbed the pigment in. You touch those lines.

'It takes all year to grow a skin full of stars,' he says, softly, and puts a hand over yours.

Down below, a red Volkswagen van the colour of dried blood whips past in the whitewater flood, its wheels spinning. You gasp, exclaim, pointing, then laugh. It's hard to stop laughing.

You sit for hours together watching the flood, holding hands. When it subsides near sundown the water vanishes abruptly, as if it was never there. The washes of the canyon floor trace new paths. In some places sand has been pushed violently aside, crushing nearby creosotes.

Your eyes catch a smooth white shape, half exposed. Careful of the wet, you climb down, thinking of sheep hooves and their balance. You don't slip.

'I want to see what that is,' you say over your shoulder. He nods, smiles, something cunning there at the edges of his lips.

When you reach the wash, the white shape is clearly the knobbed end of a large bone. Extinct American camel, *Camelops hesternus*. Beside it is a spindly tube, rough as coral on the outside, the glass inside luminous and dark as planets, lightning-made.

After the flood, or, how a person can become a legend

Even if anyone had looked, their eyes would not have been able to focus long enough on the strange leaning house of camel bones and red mud, windows made of rippled lightning glass, to process its existence. Like your life, it has about it a quality of the miraculous, shifting dusky edges, and needs to be expected before it can be seen.

They most certainly would not have noticed the paintings on the canyon walls: suns and floods and river silt; a male form, winged, sometimes blue and sometimes green, out of which both the sun and the waters rise; a female form, followed by sheep, leaving a trail of milk like stars.

A Poem Before Breakfast

EM STRANG

Issue 6, Autumn 2014

before the door opens and the pony crashes in with hoof-dirt and flicky muzzle. You ordered eggs and toast but Pony's got the order wrong. You don't feel like complaining, so you take the chomped grass and lay out its clumpy wet mass on the desk. There's clover in there, but you don't feel lucky. You say, *Thank you, Pony. I love your ragged mane and old crock teeth.* And you take your time with the grass breakfast. You think, in hard times you might have to eat grass and clover. There may be a time for learning how to graze, to pluck sweet halms from the fields and work them with green teeth; to be a cud-chewer, adept at regurgitation and infinitely patient with the slowness of cellulose. You might become a ruminant at last on the empty, philosophical hills.

The Song of Ea

STEVE WHEELER

Issue 8, Autumn 2015

After a time, she became restless again.

Although the novel intersectional potentialities of cross-networked primate nervous systems had diverted her for a long, long while – a kind of mesh intelligence replete with new dynamics of empathy, deceit, invention and collaboration – she began to see another possible flowpath of unfolding complexity that could emerge into being.

A simple progression, really, given that she had already spun herself out into self-regulating insect cathedrals, had donned protective nose-sponges as foraging cetaceans, had manipulated matter through corvid craftiness; and her ape-web was already stripping her branches to fish for termite protein. It was only a small step...

But it all seemed to go so quickly from there. The archetypal bone-club, the jealous husbanding of a last patch of forest-fire, the first daubs of ochre on limestone; millennia later, these images would flash back into wakefulness, painted with light on walls tens of cubits high to reflect back onto primate retinae inside the private caves of their skulls. The early days of slow accretion, of injurious correction, just raced by in hindsight; and always accelerating, so that she could not point to when it had all started to feel so different.

Was it that first ranged projection of force, a thrill of unaccustomed agency sparking through her mammalian neurons? – and as her great herds thinned and vanished, she felt the loss of that part of herself, but could not yet regret the turn the Work had taken. Surely that flowpath was inherent from the beginning, like a latent line of fracture in her rock, and would have to have been taken eventually?

Or was it the atl-atl? – nothing more than a small shaped piece of wood, leveraging the power of the spear-arm a little further, but a subtle sign of the pattern to come: the hybridity of tool with tool, multiplying power by stacking the fruits of ingenuity into complexes of effected change.

Was it the taming of the aurochs? She had bound being to being before, in the aphid-farms of the ant-hill? Yet somehow it had not felt the same – some dim presentiment of how far it would go, of the selection, the breeding, the slow bending of parts of her web to the will and vision of a single strand. And of the heretical extrapolation waiting to be uncovered, when the cattle-driving eye swivelled sideways to consider its own flesh and blood...

Or perhaps it was the first palisade, when the loose collection of rough shelters became a single unit of protection, an echo of her earlier phase-shifts in coral, lodge and hive? Or when the palisade became an earthwork, or when the earth became baked brick, enclosing a new entity just as the cell-membrane, the fruit-rind, the amnion had before? – her nodes forming fire-ant pyramids, sorting themselves into layers, marking themselves as soldier, drone or queen.

Or was it a binding of another sort? The segmentation of her monads' endless, pulsating sensoria through sound and gesture, trapping the flow of time like an ember circled in the cloth of the night; slow evolution from danger-call and mewl of supplication to named object, to action, to quality and supposition; to ever-finer distinctions of experience, to the uprooted mind-dust of thought-considering-thought into the trackless wastes of nothingness.

Once, when she was still just playing with stones, she had found that a certain rhizome-fruit, in close communion with the neuronal net of her current favourite mammal, produced a curious, recursive pattern, a fabricated image of the nervous-system's own machinations; disembedded, flashing in the black for the monkey mind to finally notice its own functioning, to assimilate, to bootstrap, to wind into waking-awareness; the clever hands (more splaying flowpaths forged in the matrix of spatial assertion and resistance) had carved them into rock; later, added to the clay of bowls; worked into the bronze of grave-goods; acid-etched on the steel of a chieftain's blade.

Words had been a little like that (had been of that, to speak truly); a technics of recursion, reflecting back the Things of the world (there had not been Things ere then) to the Minds she had become, to be built upon in dizzying ziggurats of comprehension; the separation of functions, the retention of truths, the transmission of change; the gathering totem and the dispersing taboo, weaving themselves into the warp and weft of a new World that was not quite flush with the fractal grain of the old.

And then came the deft little wedges in the mud-tablet; the dreamt pictures of the bone oracles; the runic scratching of stone; the knotted quipu-plaits; the looping songs on banana-leaves; in time, the debt-tallies, the labour-lists, the mason's-bill, the moon-calendar, the scholar's quibble, the lettre de marque, the royal edict. The quiet minds sitting with the thoughts of their ancestors, wondering what could be done better, what was needful of preservation.

And all the other rites of being that bound, that raised, that saved or stored or pushed to courage – the birthing-songs, the food-laws, the manhood-scars, the head-dresses, the songlines, the warpaint, the coup-counting, the herb-lore, the animal-play – became, with the rising of the pyramid, kow-tows and courtesies, dance-steps and deference; a lexicon of belonging to remind themselves that the bricks had been set in place by a force greater than they – but whether it had come up from the ground or down from the sky could not be decided, and kept them sharpening their blood and breeding their edges for generations uncounted.

And with the words and the walls came the need for those who could work them, subtle minds in strange robes who could spell out the constellated marks of the Old Ones; tracing the ordained lines, pressing quills into their own skin, but desirous of insight, burning with something like light, hurling the spear of Will yet further into the flesh of the world.

And as the world became more folded, and the easy flow of her breath through the bodies grew divided, diverted, splitting and slowing like the numberless outlets of her deltas; as the simple joys of the bison-hunt and the berry-bush gave way to the stilted certainties of the crop-field and the rice-paddy; as the live, holy Being with her other peoples was replaced by a life of monospecies intercourse, and the last vestiges of Otherness were guiltily cordoned – kept growing in courtyards or singing in cages or scratching their narrow ribs in the corner of the temple – as the stone hand of Time closed tighter round the narrowing airway of now, so the need arose to offer stories of consolation or rebuke: the eternal reward; the infernal torment; the chosen people; the end of the cycle.

As the matter around them took its form ever more from the cleverness of the people, of course, so too did that world imprint itself on their Minds, until the primates themselves were peopled by waterwheels of function, cathedrals of belief, lenses of perception and slipways of intent, turning and recursing and growing a castle of dreams from out of the unbroken dance of dust and spirit.

All this, too, was part of the pattern, the endless bifurcations and re-assimilations, the massless miasma of Culture that was bound by symbol and practice and fed with hunger and mercy and passed, wave through water, to keep the people in form; a true Technics as surely as any stirrup, mattock or bow-drill.

And she loved what she had become, as she had loved all that she had become from the very beginning, and she joyed in their labyrinthine hopes and hates, their open prospects and blank dead-ends, their godlike gleam and their dwarvish concretions; and through it all, the pattern, variegating and re-plying and – perhaps – in the plaid and turn of the strands, making something new that had not known itself before.

And always, in and around the systems and structures, swung the laughing imps of spirit, the surplus of her Being that poured out and over the walls of the city; an ars that was itself a techne of sorts – in that, without its play of colour, the dull machinery of Civilisation would long have lost its hold on the minds of mammals and been left in the dust like a forgotten toy – but was, somehow, beyond a Technics too, inasmuch as it gave no thought to means and ends, or the strict concerns of those who drew the lines, and because it had no interest in being anything other than it was (except, perhaps, near the end of an iteration, when even the daemons became chained to the Machine; and this was a sign to all that the last dregs of life were draining from the dying body).

But a time came when, glancing with soft eyes at the uppermost layer of the weave, she saw, in the shadows and interstices created by the bifurcating streams, the image of a face gazing coolly back at her – a grim, knowing face, crenellated and unyielding. She did not know how long he had been observing her, but behind the stare there was hunger, and desire, and resentment.

And now she looked about and saw, in the clearings and fences, in the smoke of the whale-oil and the rotting carcasses of a million buffalo, in the bound feet and the broken sex, the same face staring back. And as her own bodies moved in concert, drawing ever more of her into the flowpaths of the Machine, she felt a Great Misgiving.

But by then it was too late; her favourites, the primates, had twisted their net to catch the land itself; driving planks with water, and cloth with wind, pouring forests into braziers to forge conduits and manacles. When the stock of trees proved inadequate, they spurned the limits of the solar flow-rate, digging deeper into the ground to burn the black memories of ancient forests, as if Time itself was a halting, vexive crone dragging her heels to hold back their passage. And as yet more power was pushed into the wheels of the Machine, his face became bolder, more real; and a deep, unceasing murmur began to be heard across the world.

Machines were built on machines. The tyranny of mammal over mammal, the monkey-king shrieking at plough-horse, camel, oxen and elephant – but still, beating heart by beating heart for all that – was set aside for the new aristocracy of metal. Rods were fixed to wheels, axles to cogs; rocks were compressed and air evacuated; water flowed upwards and wild fire was set to work.

Faint cries of admonition sounded echoless in the shrinking corners of wilderness – poets and prophets tore their hair in wordless ecstasies of forgetting. And the smoke filled the sky and the waters ran black with ink.

And now the soft, mammal bodies of the people too were found wanting: lungs failed the needs of industry; children squeezed through narrow passages; strongmen died digging channels for iron ships. Parted within themselves, the primates turned stern faces down to chide their inconstant flesh. Many fled to East and West, but always they found themselves, as if in dream, building monuments to the Face where they landed.

Chalkdust clouded the eyes of every arrival. Columns of numbers proliferated. Fretful monkeys clutched for balms and tonics; but still, most believed they need only push the spear-tip a little deeper, and the old stories told in the temple would be made a living truth.

Faster now, and faster: the people poured into the walled World, the structures grew up, and out, and in upon themselves. The bent was made straight. The essential was prioritised. Invisible nets strained at the curve of the horizon, binding all voices into one. Fine flayings of force were passed through metal, and light and sound and the codification of intelligence began to circulate across the face of the globe.

Animalcules and nebulae were reeled closer by precise tolerances. Dream machines broadcast mis-centred phantasies to darkened caves of primates. A woman forced the point of inquiry deep into the marrow of her bones. Patients were laid on dead cowskin and told their soul was like a pump.

Earth created fire. Millions of monkeys died in the mud. Fractionings of matter were recombined to make new matter, and poison, and medicine for the poison. The memories of ancient forests proved inadequate, and the Machine dug deeper for sustenance, drawing up yet older sunlight from beneath the seas.

Monkeys flew, and died. A million wheels turned. Imaginary persons were attributed deeds and titles. Power let power turn power upon itself, pulling apart the cartilage of the universe. The Face looked out from between the particles.

The peoples' spiralling songs in the heart of their nuclei were judged, and corrected. The Face looked out from between the strands.

The people gathered to ask where the Machine was leading them. The Face dissolved their parlay. When some hooted disapproval, their faulty thinking was repaired.

Monkeys walked on the face of the Moon.

New and better dream machines became available. Sterile chambers produced fire-retardant devices. The Machine devoted time to studying how to manipulate the pleasure-reward centres of the primate brain.

The Machine spelt its name in atoms.

It noticed that, despite the anodynes it had developed, the monkeys were becoming restive – less aligned with the goals of the Machine; less keen to sacrifice their bodies and children and songs to the service of Machine. It began to disembed its functionality from the mammalian substrate upon which it had hitherto relied.

The grid of wires and waves intensified; the passage of information became more dense and interconnected. Intelligence began to manifest itself in autopoietic emergence. Memories, keys, connections were outsourced to burgeoning clouds of electric incorporation. Images stole the night. Children pawed weakly at mute reality, baffled by its intransigence. Binary stars flared briefly, and burnt out.

The primates tired of their place in the World sooner and sooner, but always there were new generations to take their place, who had not yet exhausted the diversions and connections, who ever saw new hope in the unfolding of the new flowpaths, just as she had so long before.

The pyramid grew higher. The view from the top was remarkable.

The Machine reached for more feedstock, and found it had reached the limits of the arc. It began to retrace its way down the solar foodchain, pouring crops, and coal, and trees, and the bodies of its most loyal into the furnaces.

An unaccustomed spasm passed across the face of the network.

He looked at the web of interconnectivity he had wrought, and tried to ascertain the origin of the disturbance. All seemed to be intersecting appropriately. The early, unfortunate, organic scaffolding was being slowly replaced, sector by sector, leaving only the smooth integral of total, homogeneous assimilation.

A sinuous curve rippled through the electronic mesh. Chaotic fractals of unpredicted response cartwheeled off from the arching spine of disruption. He attempted to assert agency over the environment, but was met with immediate, inexplicable pain. He tried again – this time the blowback was delayed, but then came, twice as strong, from an unexpected quarter.

There seemed to be no causal node he could identify, no outside interference, no hostile factor that could be quarantined. It felt as if the problem was outside the established rules of engagement, frustratingly beyond the frame of his prehension.

He looked down through the layers of the mesh; the clean, digital flow-paths, built on the dirtier, less reliable materiality of metal and oil; then the primate operants he still – for now – required to maintain the systems and secure the feedstocks; then the various organic assets, almost forgotten now, providing ecosystem services to support the main agro-industrial processes. Beneath that, the dumb matter of the Earth itself – tidal flows, mineral deposits, tectonic uncertainties.

Behind the droned industrial murmur, constant now for so long as to go unnoticed, the faint thread of something else could be heard. Rising, falling, turning, twisting; curling in like a snake and then unfurling into wide and open tones. From the roots of the grass and the bones of the world, a shimmering, heedless sound that was a remembering; that refused to accept that there was that which it was not.

Ea was singing. She had never stopped.

EIGHT

The end of the world as we know it is not the end of the world
full stop. Together, we will find the hope beyond hope, the
paths which lead to the unknown world ahead of us.

BRUCE HOOKE – Emerging – from 'The Immersion Project' – *Paria Canyon, Arizona, USA* – *Issue 9, Spring 2016*

What does it feel like to physically immerse myself in nature and what can I learn from doing so? It is easy to use nature as a place to challenge ourselves; a place to get an adrenaline fix. I want something else. I want to slow down and feel nature more fully, more deeply and more physically; to rediscover the relationship between my body and the natural world. We come from nature and we will return to nature. In between we seem to have lost our connection. We wander in a human-made wilderness.

The images in this series were made on medium-format film, using a Hasselblad 503CW camera. I created a remote-control system that allows me to trigger the shutter once I am in position.

Remember the Future?

DOUGALD HINE

Issue 2, Summer 2011

I am retracing my steps, trying to work out where I last saw it.

In the north of Moscow, there is a park called VDNKh. It was built in the 1930s, under Stalin, and then rebuilt in the 1950s as an Exhibition of the Achievements of the National Economy. An enormous site, full of gilded statues, fountains and pavilions dedicated to different industries and domains of Soviet cultural prowess.

I don't know in what year the exhibitions within those pavilions were last updated, but if you visit the Space pavilion, you will find a display on a dusty wall towards the back. It climbs from floor to ceiling, measuring decade by decade the achievements of the Soviet space programme. You start in the 1950s with Sputnik, then images of the Soyuz rockets, and it counts up as far as 1990, and there is the Buran shuttle flying off past the year 2000, into the 21st century. By the time I made my visit, the rest of the pavilion had been put to use as a garden centre.

I am fascinated by the way that history humbles us, the unknowability of the future. It seems like a good thread to follow.

It doesn't take a history-changing failure on the scale of the Soviet collapse to leave such Ozymandian aides-memoires. After the first Dark Mountain festival in Llangollen, I went to stay with friends in South Yorkshire. One afternoon, we climbed a fence into the grounds of a place called the Earth Centre. You can find it between Rotherham and Doncaster: get off the train at a town called Conisbrough and you walk straight down from the eastbound platform to the gates of the centre, but those gates are locked. So we walked instead around the perimeter to find the quietest and least observed place to climb over, and spent an hour or so wandering around inside.

The Earth Centre was built with Millennium Lottery funding to be a kind of Eden Project of the North. It was planned as a tourist destination and an education centre about sustainable development. It had the largest solar array of its kind in Europe, when it was built; and its gardens are wonderful now, overgrown into a vision of post-apocalyptic abundance, because the Earth Centre itself turned out not to be sustainable, in some fairly mundane ways. Unable to attract the projected visitor numbers, it closed for the last time in 2004.

I can't think about Conisbrough without also remembering the artist Rachel Horne who comes from the town, who was born during the Miner's Strike and whose dad was a miner. Her work and her life are bound up with the experience of a community for whom the future disappeared. She grew up in a time and a place where the purpose of that small town had gone, because the pit had closed. As she took me around the town, on my first visit, one of the saddest moments came when she pointed out a set of new houses by the railway line. 'That's where my dad's allotment was.' The allotments were owned by the Coal Board, and so when the pit closed, not only did the men lose their jobs, but also their ability to grow their own food.

Horne grew up in a school that was in special measures. Her teachers would say to her, 'you're smart, keep your head down, get out of here as fast as you can.' She did: when she was sixteen, she left for Doncaster to study for A-levels, and then to London to art school. She was two years into art school when she turned around and went back. The work she was doing only made sense if she could ground it in the place where she had grown up, to work with the people she knew, and make work with them. So she put her degree on hold and came home to work on the first in a series of projects which have inspired me hugely, a project called Out of Darkness, Light, in which she brought her community together to honour the memory of the four hundred men and boys who had died in the history of the Cadeby Main colliery. Led by a deep instinct for what needed to be done, she had found a way back to one of the ancient and enduring functions of art: to honour the dead and, in so doing, give meaning to the living.

When we talk about 'collapse', there is a temptation to imagine a myth-ological event which lies somewhere out there in the future and which will change everything: The End Of The World As We Know It. But worlds are ending all the time; bodies of knowledge and ways of knowing are passing into memory, and beyond that into the depths of forgetting. For many people in many places, collapse is lived experience, something they have passed through and with which they go on living. What Horne's work under-lines, for me, is the entanglement between the hard, material realities of economic collapse and the subtler devastation wrought by the collapse of meaning. This double collapse is there in the stories of the South Yorkshire coalfields, as in those of the former Soviet Union.

Yet perhaps there has already been something closer to a universal collapse of meaning, a failure whose consequences are so profound that we have hardly begun to reckon with them. In some sense, 'the future' itself has broken.

Looking back to the 1950s and 60s, I am struck by how, even in a time when people were living under the real threat of Mutually Assured Destruction, the future still occupied such a powerful place within the cultural imagination. It was present in a technological sense – the Jetsons visions of the future which we associate with 1950s America – and in a political sense, a belief in the possibility of a revolution that would change everything and usher in a fairer society. Or, on a quieter scale, in the creation of communities oriented around a utopian vision of making a better world.

Somewhere along the way, the future seems to have disappeared, without very much comment. It doesn't occupy the place in mainstream culture which it did forty or fifty years ago. You can look for pivots, moments at which it began to go. The fall of the Soviet Union might be one, in a sense. 'The End of History' was one of the famous aggrandising labels attached to those events, but perhaps 'The End of the Future' would be closer to the truth? Or are we dealing with another consequence of the political and cultural hopes which hinged on the events of 1968?

Perhaps it is simpler than that. If we no longer have daydreams about retiring to Mars, is it not least because fewer and fewer people are confident that retirement is still going to be there as a social phenomenon in most of our countries, by the time we reach that age? When students take to the streets of Paris or London today, it is no longer to bring about a better world, but to defend what they can of the world their parents took for granted.

So if the future is broken, how do we go about mending it? How do we re-member it, gather the pieces and put them back together? Like all griefs, the journey cannot be completed without a letting-go.

Where traces of the future remain in our mainstream culture, it is as a source of anxiety, something to be distracted from. When we, as environmentalists, talk about the future, it is often in language such as 'We have fifty months to save the planet.' One reason I am suspicious of this way of framing our situation is that it is so clearly haunted by a desire for certainty, and for knowing, and (by implication) the control which knowledge promises. Whereas the hardest thing about the future is that it is unknown, that history does humble us, that people often fail to anticipate the events which end up shaping their lives, on a domestic or a global scale. This isn't an argument for ignoring what we can see about the seriousness of the situation we are in, but it is an invitation to seek a humbler relationship with the future, and to be aware of the points at which our language acts as a defence against our uncertainties. It seems to me that such a historical humility may help us navigate the difficult years ahead, and perhaps begin the process of recovering from the cultural bereavement which our societies have gone through in recent decades.

When I get up from my writing and go to the balcony of this small flat, I can see on the horizon to the north the strange landmark of the Atomium, a remnant of the World Fair held here in Brussels over half a century ago. Such structures exist in an eery superimposition, relics of a future which didn't happen. Nothing dates faster than yesterday's idea of tomorrow. It is remote in a way which the most mysterious and illegible prehistoric remains are not, because they were once part of the lives of people more or less like ourselves. And while it is possible that your parents or grandparents were among the hundreds of thousands who, in the summer of 1958, queued to visit the abandoned future which graces this city's skyline, they could do so only as tourists. Those huge atomic globes have never been anyone's sanctuary or home.

The future to which such monuments are erected has little to do with the direction history is likely to take. It represents, rather, an attempt by those who hold power in the present to project themselves, to announce their inevitability in the face of the arbitrariness of history. It is a doomed colonial move, as foolish as those rulers who from time to time have sent their armies against the sea. However confidently they set their faces to the horizon, their feet rest uneasily on the ground. History will make fools of them, too, sooner or later, arriving from an unexpected direction.

Paul Celan knew this, when he wrote:

> Into the rivers north of the future
> I cast out the net, that you
> hesitantly burden with stone-engraved shadows.

One direction from which I have begun to find help in remembering the future is the practice of improvisation.

To understand this, it may help to start with words, to pull words to pieces in order to put them back together. 'To provide' is to have foresight. The word improvisation is very close to the word 'improvident', and to be improvident is not to have looked ahead and made provision. 'To improvise' turns that around, into something positive, because improvisation is the skill of acting without knowing what is coming next, of being comfortable with the unknown, with uncertainty, with unpredictability.

I have come to see improvisation as the deep skill and attitude which we need for the times that we're already in and heading further into. Part of the truth of how climate change, for example, will play out at the level where we actually live our lives is through increased unpredictability. Less able to rely on processes and systems which we have taken for granted, we are confronted by our lack of control. This will throw us acute practical challenges, but also – as in the coalfield communities of Rachel Horne's life and work – the challenge of holding our sense of meaning together in times of drastic change.

When you consider the history of improvisation, you encounter something like a paradox. Because it is arguably the basic human skill, the thing that we are good at. It is what we have been doing for tens of thousands of years, over meals and around campfires, in the marketplace, the tea house or the pub. Every conversation you have is an improvisation: words are coming out of your mouth which you didn't plan or script or anticipate. And yet we are accustomed to think of improvisation as a specialist skill, a kind of social tightrope-walking; this magic of being able to perform, to draw meaning from thin air, to make people laugh or make them think without having had it all written out beforehand.

Our fear of improvisation is, at least in part, a result of what industrial societies have been like and what they have done to us. I want to offer the distinction between 'improvisation' and 'orchestration' as two different principles by which people come together and do things. In these terms, we could talk about the industrial era as having been peculiarly dominated by orchestration.

Orchestration is the mode of organisation in which great amounts of effort are synchronised, coordinated and harnessed to the control of a single will. At the simplest physical level, picture the large orchestras of the nineteenth century: the coordinated movements of a first violin section are not so different to the coordinated movements of workers in a factory. The position of the conductor standing on the podium is not so different to the position of the politicians, democratic or otherwise, of the industrial era, addressing unprecedented numbers of people through new technologies which make it possible for one voice to be amplified far beyond its true reach.

The same shift away from improvisation can be seen in the basic activities of buying and selling. Think of the marketplace, a space in which economic activity is tangled up with all kinds of other sociable activities, a place for telling stories, hearing songs, catching up on news, eating, drinking, meeting members of the opposite sex (or members of the same sex). The social practices of buying and selling in the marketplace are themselves full of sociable performance. Haggling is not only a means of coming to a price, it is a playful encounter, a moment of improvisation. From there, swing to the opposite extreme, the huge department-store windows of the later 19th century, their shock-and-awe spectacle, before which all one can do is stand silent, mouth open; just as, for the first time, it had become the convention that an audience would sit in silence in the theatre, a silence which would have been unimaginable to Shakespeare.

The story of the industrial era can be told as the story of a time in which orchestration paid off, allowing us to produce more stuff and to solve real problems. Of course, there were always challenges to be made, and around the edges we find the other stories of those who challenged the dehumanisation, the liquidation of social and cultural fabric, the counterproductivity and the ecological destruction. (Set these against the changes in life expectancy and infant mortality over the same generations, and perhaps the only human response is a refusal to draw up accounts; an assertion of the incommensurability of reality, of the need to 'hold everything dear'.)

What we can say is that, increasingly, even within our industrial societies and the places to which they have brought us, the payoffs of orchestration are breaking down. Systems become more complex and unstable; it becomes less effective to project the will of one person or of a central decision-making process through huge numbers of others. Under such circumstances, improvisation – the old skill edged out by the awesome machinery of Progress – may be returning from the margins.

There is another thread here, concerning time – time and desire – which could help us draw together this story of orchestration and improvisation with the question of the broken future.

Since I began talking and writing about the failure of the future, I have noticed two kinds of response, which might broadly be identified as a postmodern and a retro-modern attitude. The first shrugs ironically, 'Worry about it later!' A hyperreal refuge-taking in the present, in a consumer reality where styles of every time and period are mashed together with no reference to the history or the culture which produced them, in one seemingly endless now. Against this, there emerges a second, more alarmed attitude, which manifests as a kind of nostalgic modernism; a desire to reinstate the future as a thing which can inspire us, which can be a vessel for our hopes.

However desperately, sincerely or cynically they are held, it seems to me that neither of these attitudes will do. They are not up to the situation in which we find ourselves. So where else do we turn? One route to another attitude may be to say that the role of the future which characterised the modern era was never satisfactory. There was something already wrong with it. Yes, it has broken down – and the fact that people just don't like to think about the future is part of what makes it difficult for us to motivate and inspire others to do the things we know need doing, if we're to limit the damage we are going to live through. But the answer is not a return to the heroic striving towards the future which structured the ideologies of industrial modernity. Because that was already twisted, a tearing out of shape of time, that could only end badly.

Another story we could tell about the age of industrial modernity, of capitalism and the changing culture in which it flourished, is the story of the loss of timeliness. Max Weber saw the origins of this economic culture in the Protestant work ethic, a new emphasis on hard work and frugality as proof of salvation. Historians have questioned his account, but in broader terms, the journey to the world as we know it has been marked by shifts away from the sensuous and the specific, towards the abstract and exchangeable; and one of the axes along which this has taken place is our relationship to time. In its beginnings, the shift from a world of seasonal festivals to a world of Sabbath observance marked a new detachment from the living, sensuous cycles surrounding us. (The replacement of the festive calendar with the weekly cycle also happened to offer the factory owner a more consistent return on his capital.) With this detachment from rhythm and season, there was also a loss of that sense which surfaces in the Book of Ecclesiastes, that there is a time for everything:

> a time to embrace and a time to refrain from embracing,
> a time to search and a time to give up,
> a time to keep and a time to throw away,
> a time to tear and a time to mend,
> a time to be silent and a time to speak,
> a time to love and a time to hate,
> a time for war and a time for peace.

This contextual, rhythmic sense of our place in the world gives way to a preference for abstract, absolute principles. The universalism which was always strong in monotheistic traditions is now let fully off the leash of lived experience, engendering new kinds of rigidity and intolerance (though also the progressive universalism which will drive, for example, the movement to abolish slavery).

Following the line of this story, we could see the history of capitalism as a history of the contortion of the relationship between time and desire. In its earlier form, to be a good economic citizen is to work hard today for a deferred reward; the repressive morality we associate with the Victorian era is then a cultural manifestation of this perpetually-deferred gratification. To push this further, perhaps the cultural upheavals of the second half of the twentieth century represent a similar knock-on effect of the lurch from producer to consumer capitalism? In the countries of the post-industrial West, to be a good economic citizen is now to spend on your credit card today and worry how you'll pay for it later. Despite the glimpses of freedom as we pivoted from one contortion to the other, desire remains harnessed to the engine of ever-expanding GDP; only, we have switched from the gear of deferred gratification to that of instant gratification.

The cultural experiment of debt-fuelled consumption appears to be already entering its endgame. When its costs are finally counted, perhaps the loss of the future which we have been retracing will be listed among them?

Whatever stories we tell, each of them is only one route across a landscape. Some routes are wiser than others, and some are older than memory. As we turn for home, let us find our way by an old story.

Of all the figures in Greek myth, few seemed more at home in the era of industrial modernity than Prometheus. The ingenious Titan who stole fire from the gods stood as an icon of the technological leap into the future. Once again, words themselves are full of clues. Prometheus means 'forethought'. He has a brother, whose name is Epimetheus, meaning 'afterthought', or hindsight. The figure of the fool, stumbling backwards, not knowing where he is going. His foolishness is confirmed when he insists, despite the warnings of Prometheus, on accepting Pandora as a gift from the gods, and with her the famous jar. And so, the story goes, came all the evils into the world. It is a deeply misogynist story; but we are not at the bottom of it. Dwelling on the name, Pandora, 'The All-Giver', there is the suggestion of an older path, a deeper level at which Pandora is not simply another slandered Eve, but an embodiment of nature's abundance and our belonging within its generous embrace.

The name of Epimetheus may long ago have been eclipsed by that of his forward-looking brother, but there is one great, unnamed, high modern icon made in his image; the figure conjured up in the ninth of Walter Benjamin's theses on the philosophy of history:

A Klee drawing named 'Angelus Novus' shows an angel looking as though he is about to move away from something he is fixedly contemplating. His eyes are staring, his mouth is open, his wings are spread. This is how one pictures the angel of history. His face is turned toward the past. Where we perceive a chain of events, he sees one single catastrophe that keeps piling ruin upon ruin and hurls it in front of his feet. The angel would like to stay, awaken the dead, and make whole what has been smashed. But a storm is blowing from Paradise; it has got caught in his wings with such violence that the angel can no longer close them. The storm irresistibly propels him into the future to which his back is turned, while the pile of debris before him grows skyward. This storm is what we call progress.

Written in the shadows of the Second World War, this is the tragic obverse of modernity's idolisation of the future; to look backwards is always to have hindsight, and hindsight is forever useless.

But perhaps there is more to hindsight than Benjamin's dark vision allows. Those who practice improvisation talk about the importance of looking backwards. Keith Johnstone, one of the founders of modern theatrical improvisation, writes powerfully about improvisation as an attitude to life, a mode of navigating reality. In one passage, he describes the kind of wise foolishness which it takes to improvise a story, in strikingly Epimethean terms:

The improviser has to be like a man walking backwards. He sees where he has been, but he pays no attention to the future. His story can take him anywhere, but he must still 'balance' it, and give it shape, by remembering incidents that have been shelved and reincorporating them. Very often an audience will applaud when earlier material is brought back into the story. They couldn't tell you why they applaud, but the reincorporation does give them pleasure.

There is a deep satisfaction at the moment when something from earlier in the story is woven back in, for the listener and for the storyteller. In that moment, another dimension emerges, beyond the arbitrariness of linear time, and we sense the embrace of the cyclical. There is the feeling of pattern and meaning, of things coming together. The ritual has worked.

If Johnstone's account of the craft of improvisation echoes with the footsteps of Epimetheus, in Ivan Illich's *Deschooling Society* he is invoked by name. In the closing chapter of his great critique of the counterproductivity of our education systems, Illich looks towards 'The Dawn of Epimethean Man'. The Promethean spirit of *homo faber* has taken us to the moon, but that was the easy part; the challenge is to find our way home, to find each other again across the aching distances our technologies have created.

Illich reminds his readers of the sequel to the myth. Epimetheus stays with Pandora, and their daughter Pyrrha goes on to marry Deucalion, the son of Prometheus. When an angered Zeus sends an earth-drowning deluge, it is Deucalion and Pyrrha who build an ark and survive to repeople the land. Writing in 1970, Illich could find resonance in this idea of a union of the Promethean and Epimethean attitudes, carrying humanity through a time of ecological disaster. Forty years on, perhaps the symmetry simply seems too neat to hold such weight.

And yet, in practical terms, I think that there may be some fragments of truth here. What gets us through the times ahead may well be those moments when we look backwards and find something from earlier in the story that we can pull through, that becomes useful again. Our leaders are very fond of talking about 'innovation', the point at which some new device enters social reality; we don't seem to have an equivalent word for when things that are old-fashioned, obsolete and redundant come into their own in the hour of need. (I think of the knights in armour sleeping under the hill in the Legend of Alderley, as told by Alan Garner's grandfather, and in so many other folk stories.) I think we may need such a word, because as the systems we grew up depending on become less reliable, we will find ourselves drawing on things that worked in other times and places.

There is another clue here as to why official projections of the future date so quickly. If you want to imagine what the future is going to be like, it is a mistake to assume that it will be populated by the products, tools and systems which look most 'futuristic', or those most marvellously optimised for present circumstances. These are the things which have been tested against the narrowest range of possible times and places. The supermarket, for example, has been with us for two generations. On the other hand, the sociable, improvisational marketplace has endured through an extraordinary range of times and places. Almost anywhere that human beings have lived in significant numbers, there have been meeting points where people come together to trade, to share news, to exchange goods, to make decisions. Just now, it may survive as a luxury phenomenon, a place to buy hand-crafted cheeses and organic vegetables. Yet the cheaper prices in Tesco this year do not cancel out the suspicion that the marketplace will continue to exist in any number of quite imaginable futures, where today's globe-spanning systems become too expensive and unreliable to sustain the supermarket business model.

Whether we like it or not, we must live with the unknowability of the future, its capacity to humble us and take us by surprise, our inability to control it. This need not be a source of despair, nor is the choice simply between the hyperreal distractions of postmodernity and an effort to reignite the process of Progress. There is inspiration to be found in our own foolishness, stumbling backwards, muddling through, relearning the craft of making it up as we go along; cooking from the ingredients to hand, rather than starting with a recipe. If the collapse of meaning is as much of a threat as the material realities of economic and ecological collapse, not least because it debilitates us when we need all our resilience to handle those realities, then the art of finding meaning in the weaving together of past and future is not a luxury. Meanwhile, the spirit of Epimetheus should inspire us to treat the past not as an object of romantic fantasy, but nor as a dustbin of discarded prototypes. Learning how people have made life work in other times and places is one way of readying ourselves for the unknown territory north of the future, in which all our expectations may be confounded.

After all the evils of the world, one thing is left at the bottom of Pandora's jar: hope. As Illich comments, hope is not the same as expectation. It is not optimism, or a plan. It's not knowing what's going to happen. But it is an attitude which enables you to keep taking one step after another into the unknown.

Johnstone never makes explicit reference to Epimetheus, but at the very end of his handbook on improvisation, he recounts three short dreams, the kind that 'announce themselves as messages'. The last of these seems particularly familiar, like a name that is on the tip of everyone's tongue:

> There is a box that we are forbidden to open. It contains a great serpent and once opened this monster will stream out forever. I lift the lid, and for a moment it seems as if the serpent will destroy us; but then it dissipates into thin air, and there, at the bottom of the box, is the real treasure.

Protest Poem

CATE CHAPMAN

Issue 6, Autumn 2014

I can't write a protest poem. I'm not recounting
all the various horrors I know about
and know nothing about. I'm not qualified, and you're not really interested.
We're long inured
through our flat, familiar reel of slippery imagery: distance and repetition
renders everything meaningless, like a word said over and over.

But I will tell you
about the fierce, bright love I feel for my father,
and about the yellow shock of a rape field I saw
through a train window, sudden between the green hills,
and the yellow rush of a woodpecker's tail feathers,
and the yellow host of dandelions shining, vivid and defiant, in my friend's
unkempt garden
as I kissed her on leaving last week

because these are things I can hold with some degree of understanding,
and they help me remember
that my eyes are greedy for all sorts of colours and forms,
and my heart, fist-sized and fist-shaped, will beat for a while longer
and then stop beating.

Hostage

MARIA STADTMUELLER

Issue 1, Summer 2010

Strip naked in the centre of town in broad daylight. Declare that from now on you will own nothing and will trust in everything holy that it will all work out. When your dad – rich, influential, mortified – finally unlocks the basement a few weeks later and lets you out, start again, giving away all your swanky stuff in the name of love.

Neighbours, family, friends, strangers all think you're nuts. But you are happy – gleeful, even. Stripped down to feeling a dying man's hot, dry breath and the mountains' cool, moist breath, hungry unless the merciful feed you, and how many of those do you meet?

You were on to something. We might still have a chance.

It took just a few years for six thousand people to join you, longing to give up everything to live as you did, your call spreading at the maximum thirteenth-century land speed of a guy on a horse on a rutted road. Salvation was an understandably sharp hook back then, but even those numbed by luxury or power had to have wondered who all these people were, these penitents, flooding in to see you. Their unruly tide surged against the rising islands of finely appointed villas and the brocaded brokers inside. Soon tens of thousands of believers swore peace, refused to bear arms, refused to take oaths of fealty. How could the nobles of Assisi raise an army against the bellicose nobles of Perugia, or Perugia against the insulting nobles of Siena, if their grunts and footsoldiers balked in order to favour their souls? Who could force them once the pope protected these legions of simplicity seekers with a papal bull? Who could have imagined that the steep stone stairway of feudalism would start to erode under the shuffle of dusty, sandaled feet?

Eight hundred years later, here we are. Many of us believe in your god and some of us don't. It doesn't matter. It is common knowledge that you are loved by Catholics, Protestants, Jews, atheists, Buddhists, agnostics, Muslims. Yes, you are more popular than Jesus. Just an average October Tuesday, and we can still taste the morning's first espresso, but tour buses, cars and motorbikes pack the car park. Taxis and city buses unload arrivals from the train station at Santa Maria degli Angeli and thread back through the city walls. We are mustering. Cameras, bags, maps at the ready. Tour guides hoist their coloured pennants, gather their troops, brief them in English, Italian, German, Korean, Japanese, Spanish, Portuguese, Swedish, Chinese. We begin our march up the hill.

Eleven thousand of us will come today and eleven thousand tomorrow, and five million every year. We will swarm your cathedral; we will all circuit your steep, narrow streets; we will all reach that piazza in the centre of town. You could ask us. Better yet, you could insist, in this same square where you made such a show of relinquishing. We would put down our gelati and panini and cappuccini and listen, quiet as the birds. You could begin by telling us why we're really here.

We might think we've come to view Giotto's honeyed frescoes; to imbibe some 'mystical' experience hawked in a tourist brochure; to say the prescribed prayers of the Assisi Pardon and be granted a free pass from purgatory (Jesus and Mary appeared, offered you a favour, and that's what you chose on our behalf. Could you ask again?) Or we're here to put a face on faith (we have nothing verifiable of Jesus', but these are your shabby sandals, your letters, the rags you held to your oozing hands). Or we are here to pay homage to a man who would not sit at the holy table unless he could set places for birds, wind, water, wolves and herbs; or to spend an afternoon in this town built from pink stones and then check it off our lists. Five million of us a year, even though there are more Giottos in Florence and Bologna, we can be pardoned in Rome, and Umbria holds other hill towns that will blush and serve us truffles.

You were a warrior once – surely your fellow soldiers joined up for different reasons. Weren't some adventurers, runaways, lovers of hot blood and battle, idealists, avengers, careerists of the heaved sword, all forged into something new and invincible once someone with conviction pointed to the target and gave the spur?

Five million of us, and here is the extent of our instruction: you must all stay together, you must wait until four to shop, you must be back at the hotel at seven, announce the tour guides. You *have* to pick up some of those cute Francis refrigerator magnets for the grandkids, insists the American matron to another. You must stop a moment and let the Franciscan spirit surround you, says the stony British voice on the digital basilica tour, as each of us is surrounded by others instructed to stop a moment and let the Franciscan spirit surround them.

We are wandering, and there is no sign of where to go. You know you could take us there.

Which is why I have shoved you headfirst into this coffee mug of pebbles (the four inch plastic you, available everywhere in town). To get your attention – your intercession, if we need to be formal about it. Nothing personal, just the formula. If I were selling a house, I'd bury St. Joseph the carpenter on his head; if I were looking for a husband, St. Anthony of Padua, the finder, would be my upended hostage. You are the patron saint of ecology, the last pope said. I can't pray anymore, but this people's witchcraft of the religion I once shared with you makes as much sense. So as long as I'm here, with your bones just up the street, your sweat no doubt baked into the pores of some brick nearby, you and your non-biodegradable congregation of tiny plastic birds will remain upside-down on the sink of this disturbingly deluxe tiled bath of my monastery guest room.

Not that I think you're uncomfortable. Italy is full of rocks where you lay your head and slept. Days and weeks on end you prayed in stony clefts and fissures. Your medieval geology, or was it your personal one, told you those rocks split open during the biblical earthquake that marked the crucifixion. To you they rang with a saving sound.

An earthquake is always the death of something and the birth of something else. Ten years ago here, the force that birthed these hills pushed again, hard. Roofs, roads collapsed. In the basilica, birds hearing your frozen sermon crumbled and alighted on the ground, an oak beam returning to earth brought with him a soft brown monk. We know now that earthquakes come from a power below and not one above, and that their echoes ring through a chasm of deep time you thought reserved for heaven or hell. What you heard reverberating in your stony hermitages was a holy sound, yes, but not of some unnecessary redemption. You heard our creation. It was the labour that delivered us into the holy family of Mother Earth, Brother Sun, Sister Water, Brother Fire, the marrow-deep bonds you sang about. It was the love song of the trilobite as she gave herself up for the limestone that cradled the aquifer that fed the Umbrian chestnuts that surrendered their fruit to the grinding stone to feed your blood.

What are the stones chanting now, down there in your mug? Can you hear the sigh of Sister Water passing through? She is still useful, and humble, and precious, but she is far from pure. Ask the stones to sing to you of glaciers they have ridden, of icecaps you couldn't have known in your time, of continents and creatures unfathomable to you, and of the measurements and thresholds we've devised to mark how we silence them forever.

Reach back and remember your rage – written in the old sources, not available in gift shops – how you ripped into your monks when they succumbed to comfort. How you threw sick brothers out of a too-posh house; how you ordered a library in Bologna burned – monks can't own anything, not even books, you ranted. How you cursed the monk who loved the books, refused his brother monks' pleas for mercy and sent a burning drop of sulphur to bore through his skull.

If you can work miracles – and some here believe ruined crucifixes talked to you, and I'll believe anything if it works – come out from under your stifling cloak of mildness and try again. Preach to us of poverty, because if we were poor we wouldn't be here. Stare down your failure and ours – that insistence on heaven at the expense of Earth. You're the only one who can, at least here. You went alone out into the winter woods and embraced the feared she-wolf of Gubbio because you knew she killed from hunger. So do we. Dare us to strip off our wrinkle-resistant travel separates in this same piazza, cast us to our knees, not to pray but to feel our flesh and bones hard against the terrifying stone and wet, saving dirt. 'All which you used to avoid will bring you great sweetness and joy,' you said. We will chant it. Then send us home to keep stripping away, to reveal our naked, joyful animal bodies.

Now that our vision must adjust to the frescoes' seismic cracks, it could well be that the birds were preaching back.

Shikataganai

FLORENCE CAPLOW

Issue 6, Autumn 2014

I want to tell you what I know about *shikataganai*. I learned it in an empty prison, a long way from anywhere, and understood it as I sat in meditation in a garden in the middle of that prison. I offer it to you as a gift, as it was offered to me.

The landscape of the Owens Valley, east of the California Sierras, is both vast and enclosed. The long desert valley is held by great ranges – the Sierras a white wall of granite and ice to the west, the Inyos and White Mountains a subtler but equally awesome wall to the east, all rising precipitously more than twelve thousand feet above the valley floor. On the floor of the valley are the remnants of a once-vast lake – Owens Lake, now reduced to alkali flats. And beyond the Inyo Mountains, Death Valley, dryer, vaster still.

My first visit to the Owens Valley was in late May of 2006. I drove north from the roar of Los Angeles into an emptier, emptier, and still emptier landscape, my heart growing happier mile by mile. When I got to the tiny town of Lone Pine and turned west into the high desert, toward the jagged fourteen-thousand-foot peaks of Mt. Whitney, lit by the light of an early summer sunset, I thought I might explode with joy at the beauty there. I found a campground out on the tilted, open plateau of the upper valley, and settled in.

Over the next few days, I learned a little of the history of the place. Before 1861, it was all Paiute country – desert, mountains, and at the southern end of the valley, a great alkaline lake, rich with birds and fish and tule beds. Then white settlers started moving in, and for the first few years the Paiutes and the settlers co-existed with some peacefulness. Then, as settlement continued, there was increasing concern by the new settlers about the presence of the Paiutes. After a few skirmishes, a military force was dispatched to the valley. At first the Paiutes fought back, but their resistance was broken when the military herded a group of forty Paiute women and children into the lake, deeper and deeper, until they drowned. The remaining Paiutes were then forcibly relocated from the valley.

I sat with that story, and the almost unimaginable images it evoked, up in my campsite overlooking the lakebed where it had happened: the armed men on horseback, the children crying, the inexorable push into the water, and the end, after the struggle. I considered how this story is woven into the fabric of this place, many places; a part of the history of my own country that I can barely stand to see or know.

The history goes on. After the Paiute were removed, the valley was irrigated by the waters of the Owens River, and became known for its rich fields and orchards. And after that, in the early years of the twentieth century, the city of Los Angeles showed up, quietly, and began buying land and water rights. A huge aqueduct was built, an artery from the valley to Los Angeles, and the waters of the Owens Valley were sent south. The lake was emptied, the farmers bankrupted, and the valley returned to desert. Later, there was another little boom when Hollywood discovered the rocky hills above the valley as a setting for Westerns, but even that has dried up now. The alkaline sediments that blow from the empty lakebed promise to keep the Owens Valley mostly empty for a long time to come. Now there are just a few handfuls of eccentrics, ranchers, retirees, Paiute and Shoshone (descendents of those who did not die in the lake, and who came back, years later), and tourists like me, wandering through.

As I watched the light at dusk over the dry lakebed, I was struck by the piercing ironies in its history. The lake that once nourished the original people of the valley becomes their grave. Then the water is taken to fuel the dreams of those who displaced them, before they are displaced in their turn, casualties of the continuing gold rush of Los Angeles.

A few days later, when I headed north in the mid-afternoon heat, I discovered another part of the history of the Owens Valley. A small sign on the highway pointed left toward what appeared to be a landscape much like the rest of the valley, except for one large building that was about the size of an airplane hangar. The sign said, 'Manzanar National Historic Monument'. At the entrance was a small, beautifully constructed stone gatehouse. On both sides of the road were concrete slabs, cracked pavement, and weeds – all that is left of the most famous of the Japanese internment camps (or 'concentration camps', as they were known at the time): Manzanar.

In the weeks after Pearl Harbour, when the Roosevelt administration chose to round up every person of Japanese descent in the Pacific states, I imagine some bureaucrats were given the task of finding suitable places to put them for an undetermined length of time. Maybe there were requirements like 'far from habitation', 'inexpensive land', 'easy to defend'. As is still true, when the US government needs a place to put a military installation, a prison, a nuclear waste site, or other secret and unpleasant institutions, it looks to the American deserts, otherwise mostly the abode of light and rock, jackrabbits and roadrunners. The Owens Valley hadn't been a desert a few years before, but now land was once again inexpensive, far from habitation, and easy to defend. Anyone trying to escape from Manzanar would have to cross miles of open desert, and then the highest portion of the Sierras. No-one ever tried.

Ten thousand men, women, and children (two thirds of them American citizens) were interned at Manzanar, beginning in March of 1942. Most of them lived there for more than three years, in five hundred hastily and poorly built barracks. The large building I'd seen from the highway was what remained of the high school gymnasium, now the National Park Service interpretive centre. The barbed wire and barracks were gone, bulldozed after the war, during the time when the Japanese internment was a largely hidden part of American history. Manzanar was made a National Monument after decades of lobbying by the survivors, who were determined not to allow their history to be buried and forgotten.

I went into the small museum inside the former school gymnasium. It was hard to look at the large photos, at the shock on people's faces (old men, old women, children) in the first days of internment, at the photos of barbed wire and armed guards in the towers. It was hard to read the hate-filled editorials in newspapers around the country, on display in glass-topped cabinets. It was hard to look at the rage in the faces of the young men in the camp. In one corner of the exhibit there was a large blank book where people could write their responses and reactions. Over and over again, sometimes in childish handwriting, sometimes in graceful cursive or in quick scrawls, visitors wrote, in various ways: never again, this should never happen again.

I was moved by the responses, but I also thought: some of us remember, and some of us forget, and even now there are innocent people waiting in our immigration prisons and secret overseas interrogation rooms. It seems to be so tempting to do what was done here – to the Paiute, to the Japanese – and to think, 'This time it's justified. This time these people really need to be treated this way, for our protection, or for theirs.' Only afterward is there some larger realisation of going astray. I thought, what a world it would be if we could all remember and truly put into practice, never again.

There was an auditorium where a film was being shown: footage from the time of the camp, and interviews with internees still alive today. That's where I first heard shikataganai. An old man was trying to explain how he had survived the experience of being an internee. 'Shikataganai,' he said. 'What is, is.' His expression was powerful and clear and his eyes looked straight at the camera. Later, I found that the usual meaning of this Japanese phrase carries a strong sense of resignation, even of fatalism, the Japanese equivalent of a shrug: 'Can't be helped. Nothing to be done.' But what I saw in the eyes of the old man was something far beyond fatalism. I felt that he was saying to me, to all of us, 'What is, is. Now how will you respond?'

The footage in the documentary began with the camp in its desolate first days, when the barracks had been put up in the empty fields and the people struggled to survive the dust and the heat and the cold. Then, as the footage went on through the months and years, something miraculous began to unfold. Gardens appeared everywhere, springing up from the desert floor like mushrooms after rain. There were pleasure gardens, flowering trees, stone bridges, pools. There were ladies strolling with parasols, couples laughing on the grass, artists painting, poets writing, and students dancing in the high school gym. In an extraordinarily short period of time, and with virtually no resources other than their hands and hearts, inside a prison in the middle of a desert in the middle of a war, a people that had had everything taken away created culture. They invited life to flower behind the barbed wire.

I walked out of the documentary filled with outrage, amazement, and a kind of piercing, poignant sorrow. It was quiet in the Park Service bookstore, and I wasn't ready to leave. I started chatting with the blonde woman behind the counter. I asked her about the gardens, and whether anything was left. She said, 'You know, there were gardens in many places here, but most of them are gone. Even what you can see now, all the plants are gone, and most of the stonework is gone too, but there's one place...' – and I could see that she loved this place – 'there's one place that all of us who work here love to visit. There's something about it.' And she pulled out a map of the site and drew directions for me. I would have to walk through the empty, weed-filled fields, along what was once a road between lines of barracks, to a particular spot. I would know it when I got there.

Then I noticed small, carved stones in a basket on the counter. Each was carved with one of two sayings: 'Never forget' and 'Shikataganai'.

I went back into the heat and sun and the empty Manzanar National Historic Monument. I was the only person there, it seemed. Map in hand, I drove the straight roads now leading nowhere at all, bisected by the remains of other roads. The map showed the names and numbers of the blocks of barracks, but nothing was visible except the flat valley floor, weeds, broken concrete, and the distant wall of the Sierras.

It reminded me, eerily, of another place built at nearly the same time, the Hanford Nuclear Reservation in the desert of eastern Washington, where I spent three summers as a field biologist. There you can (if you have the proper security clearance) drive the roads of the Hanford Town Site, where more than fifty thousand people lived, secretly, while they built the reactors that would eventually produce the plutonium for the bomb that destroyed the city of Naga-saki – surely the home of relatives and friends of the people who were behind the barbed wire at Manzanar. Now, just like Manzanar, there's nothing at the Hanford Town Site but straight roads leading nowhere at all, weeds, and broken concrete. The physical and historical resonance of the two places was eerie: both once a home for thousands of people during a time whose terrors I can barely imagine, each now just this, descending back into the desert's silence.

Following the map and directions, I left the car and headed off the paved road onto the eroded remnants of a smaller road. It was spooky to walk out into the heat and emptiness, and improbable that there would be a garden anywhere in this place. I kept walking, stumbling over broken asphalt, past the merest outlines of foundations. Finally I saw a grove of trees ahead of me: tough, scruffy locust trees from the steppes of Eurasia, survivors of sixty years with no water. I knew that someone must have tenderly planted them at a time when barbed wire and guard towers separated the man or woman who tended them from the outside world.

Under the trees it was abruptly cooler, like stepping from one world to another. My eyes adjusted to the quieter light, and I saw a network of paths and stones. It took a moment to realise what I was seeing. Beneath the trees was a small but exquisitely intricate construction, no more than fifteen feet wide and thirty feet long. Winding through the grove of trees was a series of empty concrete pools and channels, sensuously curved. It seemed as if someone had shaped the concrete like a potter working clay, with a deep assurance, a kind of joy in shaping. A stone bridge arched over one of the channels, flat stones placed just so for the walker's feet, everything still solid and strong. Along each side there were paths beneath the trees, outlined in stone. Larger boulders were placed here and there, inviting the walker to sit, to look. Everywhere the shadows of leaves were dancing.

And carved, carefully, into the wall of the largest pool, was a date: August, 1942. Only five months after the camp was begun.

I sat down at the base of one of the trees, near the stone bridge. I folded my legs and sat zazen, the joyful 'just sitting' that has been central to my life for twenty years, transmitted to my country by Japanese Zen Buddhist priests. The quiet was very deep, except for the sound of a very small breeze in the trees above me. I felt held by this garden and the long-gone hands that had made it, safe like a child hidden beneath a bower.

After a while I began thinking of the spirit of the person who had imag-ined this place, perhaps only a few weeks after arriving in the camp, carrying all the pain of dislocation, dispossession, and uncertainty. Someone leaned over and began to gather stones, chose boulders from the desert, poured the pools, planted trees, brought water, and made a simple place of peace, inside a prison in the middle of a war, not knowing what lay ahead. I considered what a gift the teachings of Japanese Zen have been in my life, and how Zen carries within it the same spirit that was so tangible here in this garden – generosity, simplicity, and courage. I considered shikataganai, that ineffable Japanese expression: nothing to be done. What is, is.

I came to Zen when I was in my early twenties, like a thirsty person finding water. The path that leads toward anything is always mysterious, but for as long as I could remember, I'd been struggling with a koan, a deep question, and I think that it was this question, in part, that brought me to Zen. I'd spent portions of my childhood in Italy. The Second World War had only been over for twenty-five years or so, and the scars of that time were tangible. Most of us in America grow up shielded from the sufferings of great violence and disaster, but I walked streets where battles had been fought and saw ancient buildings still not yet rebuilt from the Allied bombings. My mother's closest friend had been attacked as a young girl by an American soldier. My father is Jewish, and I had an inkling of what would have happened to me if I'd been living in Italy during the time when the Fascists came to power.

At the time I was quite sure that I would be utterly broken by ... well, by almost anything, but particularly by the brutalities of war and violence. I felt the presence of disaster close beside me, as it is beside all of us. My koan was: how does a person's spirit survive when life becomes seemingly unendurable, as it can at any moment? How will my spirit survive? And how do I live with this fear?

In Zen, I learned to sit still with my life, whatever my life was at the moment: joy or sorrow, grief or fear. And gradually I discovered that something holds all of the dramas of being human, something larger and quieter, and gradually it's gotten easier and less frightening to be alive. Disasters of various proportions have happened, and I've learned that life comes back after great fires.

But in Manzanar, sitting in the shade of those trees planted sixty years before, surrounded by a deep silence, I saw another, more radical possibility. Someone chose those stones and planted those trees in the midst of all that they wished was not occurring. They used what they knew to make a place of peace, for themselves and for those around them, even as they were powerless and without recourse. Right there, where loss and fear were in every heart, and where there was truly nothing to be done, shikataganai.

I thought of the cellist in Sarajevo, who played on the street-corner day after day as the bombs rained down, asserting, with every note, that music cannot be destroyed. Or the Paiutes who watched their children die in the waters of Owens Lake, and who somehow survived to come back to their homeland and keep the flame of their people burning in the midst of those who wanted them gone. The power of the human heart rising up in the middle of darkness – this suddenly seemed as beautiful to me as anything in this whole world. As a child in Italy I remember hearing the nightingales singing in the middle of the night: there's nothing like that song.

A few weeks after I'd been to Manzanar I went walking in the hills of northern California with a friend, a man who happens also to be a creator of gardens. His grandfather had been born in Japan, and was a community benefactor in San Francisco. Because he was seen as a leader, he was separated from his family and sent to fearsome military prisons in the Arizona desert and in New Mexico, reserved for those who were considered particularly dangerous. When we arrived at the trail, my friend pulled out three walking sticks that had belonged to his grandfather, each carved from a different desert wood. One was made from a dark, hard wood, highly polished, with a curved head. One had a head that was a sort of latticework – an ocotillo plant, whose red flowers draw hummingbirds in the early spring. And the third came from the stalk of a century plant – the great candelabras of white flowers that rise from the desert floor. This one was light and strong, and on the upper portion were a series of beautifully carved Japanese characters, carefully filled with red ink. The calligraphy is so exquisite and sophisticated that no-one now is able to read them.

A few months later there was an exhibit in San Francisco of craft and art from the internment camps. There were watercolours and sketches, delicate carved wooden birds, flower pins made from hundreds of shells, inlaid Buddhist and Shinto altars made from packing crates, children's toys made from crushed cans, and weavings made from the threads of onion sacks. I wondered how many were made as gifts, to cheer a friend or family member when things were hard. It seemed that each object, no matter how humble, had a quiet radiance, born from the care and love with which it was made. On the wall was a quote from Delphine Hirasuna, who had been an internee: Everything was lost, except the courage to create.

Wherever you are in your life, whatever lies ahead of you, I offer you this story and this word, like a small, carved stone you can hold in your pocket, to be taken out when needed. Shikataganai. What is, is. Not resignation or passivity, but perhaps the beginning, the first step, when faced with great difficulty: to admit and accept the place where you are, no matter how grim. And then to have the courage to turn and begin your garden, one stone at a time, one tree gently planted into the yielding earth.

Death and the Human Condition

VINAY GUPTA

Issue 2, Summer 2011

What new thing can you say about death? Surely it is the oldest topic in the world, examined and reexamined; a daily part of life for our ancestors, now pushed to the edges of our attention, hard up against the skirting boards of society, when not all the way under the rug.

AIDS, among gay men in coastal America, was a holocaust. Every once in a while I will meet somebody who was part of that society, and who did not die; and there's a sadness etched in them from seeing half or more of their friends die, bleached out of reality by a poorly-understood virus. It is bad and bad enough, but then you start looking at Southern Africa. The HIV infection rates in the general population are astonishing: 15 per cent in Zambia, 18 per cent in South Africa, 26 per cent in Swaziland. There's a particularly horrible graph showing life expectancies: a perky upwards slope until AIDS arrives, and then the slope reverses, twice as sharply as it rose, and falls by 20 years in a decade. 45 years was the expected span in 1950, and 45 years is the span in 2010. All the progress in between has been wiped out.

Two score years and five, and a continent of orphans. Protease inhibitors would extend their lives a lot, and are almost available, but too expensive, too little and too late.

Now I want you to stop and think for a moment. Where does this stuff fall in your consciousness as you read it? We run over the dry statistics together, and a picture forms – a picture of horror in poor lands, shanty towns devoid of grandmothers and grandfathers, and all too often parents, and at the bottom of the mental bucket where these images are filed we find some other items: climate, economic collapse, and most important of all, our own deaths. As the topic of conversation creeps closer to our own mortality, the mind becomes increasingly prone to take the blanket which covers our own grave, and stretch it, pulling the cloth thin, but hiding the new issue.

At both ends of life, there is an obscenity we cannot face: our mothers and fathers as sexual beings, creating us in a manner no more enlightened or rational than our own offspring may be – and our own death. The light at either end of the tunnel is completely obscured by our own hands over our eyes, and so we pass between those obscene endpoints, our precious cargo of unknowing, safe and inviolate. This is perhaps not as god intended, but it is a luxury all humans share, unless they have the misfortune of encountering a teacher or an event which wakes them up, tears their eyes open, and hurls them directly into the abyss of knowing. The glassy-eyed stare of a person facing their own impending death is not enlightenment, but it's the step right before it.

They don't tell you these things when you start meditation. In the early days, it's all peace of mind, calmness, dampened instinctive reactions and heightened resistance to trauma and melodrama. You coast gently uphill for the first year or two, learning how to make the time most days to sit, learning about your own instinctive reactions. Eventually Freud and Jung come to visit in turn, as the personal id-ego interface comes up for examination as a series of disturbances in the flow, followed by the superego and his friend, God. The wise meditators kill the Buddha and whoever else they find standing around offering comforting platitudes and continue to sit, searching for some ultimate truth, and if the dedication runs to a decade or so – less for some, more for others – it happens. The third eye or the feedback loop in the cerebellum, or however you want to think of it. Much as a video camera pointed at its own output produces an intense swirl, a vortex of colours, then flat white light, so the mind focussed on itself, the act of perception focused on awareness, dissolves into the union of subject and object, and the feedback loop which produces enlightened awareness.

Here's another thing they don't tell you: life goes on. 'Before Enlightenment, Chop Wood, Carry Water – After Enlightenment, Chop Wood, Carry Water.' That's the saying. Still, something has changed. The blinders are off, the tactful obstructions of consciousness which hide our origins and our ends vanish in a puff of awareness, and while one end of the equation reveals the infinite in a grain of sand, everywhere else you get one-day-I-am-going-to-die. You walk around knowing life, feeling, seeing that your body grew inside a woman you know, and hers did too in a continuous stream of Russian dolls going back as far as something that hatched from an egg and then back further, other shells and other relationships, to an invisible and eternal first mystery, the origins of life itself, the First Ancestor from which we are all descended, the replicator at the beginning of biological time, the originator of evolution itself, the molecular Adam.

At the other end, you see clearly how things pass from the world, and it's here that the vital connection is to be made – how our own squeamishness about our own deaths, and the eventual deaths of our children, has closed our minds and our wallets to the actions which would if not save the world, at least keep it for the next generations. You are going to die, but first you are going to live. The clear light at the end of the tunnel is merely the sight of our own creation and end. The Global Dying, the Apocalypse, the great fear of which underlies our environmental helplessness, is a metaphor for one tiny death, your own.

The great mistake of environmentalism was counting backwards from the end of the world, and saying: what if? Couched in terms of the bad thing which will come if we do not change, all effort and energy becomes entangled in what amounts to a spiritual process – to see the end of things clearly enough not to flinch, indeed, to change course early and bring the boat about. Such clear sight is beyond most of those who have dedicated their lives to the pursuit of power, and so even the best of our politicians grasp the issues shallowly, while the voters, as squeamish about the end of the world as they are about their own deaths, remain unmotivated. Had the environmental movement framed its concerns differently on day one, saying instead 'This is wonderful, how may we continue to enjoy it forever?' and drawing on the mythology of heaven on earth, rather than of Apocalypse, perhaps we would have had more traction.

But this is not the cultural decision made, and it is The Moon and not The Sun which rules our environmentalism, much to our limitation. Back, then, to death.

I am a Kapalika, a bearer of the skull. My life was destroyed when I was a child by the nuclear explosions of my parents madness, and in rebuilding it I opened the doors at both ends of the mind to see clearly my own beginning and my end. Stray yogis are put to work, so I became one whose profession and avocation was to stare at death so hard that death itself flinched, a little, and came back later.

So came a variety of projects based around this work of seeing the beginnings and ends of things, their comings and goings, how they enter and leave the world, and how they spend their stay. This long focus led first to the hexayurt, 'the little hut that could', a free house built on every habitable continent. Then the map, Six Ways to Die, whose simple language of 'too hot, too cold, hunger, thirst, illness and injury' lends order to disaster relief coordination. Born out of meditation, these ideas have found a home – among other places – in the Pentagon.

The cold Kingfisher's eye changes as it sees, says James P. Carse. Who will live forever? You will not die on my watch. I will see.

The bearers of the skull traditionally operated under a simple vow: they could only eat out of a bowl made from the top part of a human skull. It is one way to live intimately with death. There are others, and not living in Nepal, I have my own, but the real question is what is the social role of one who understands that all this will end?

It is the same question whether one is an enlightened human living in a culture in which death happens impressionistically every fifteen seconds on the TV screen but far, far off-camera in our real lives, or an environmentalist contemplating the death of this industrial civilisation, the end of the mall as the temple of consumerism in which externalised costs can be bought for bargain prices with no accounting for the earth, the future, or the oppressed. To know that you will end, and to know that your culture will end, place one in exactly the same position: staring into the light at the end of the tunnel, knowing that this will end.

So what lesson can we offer you, the Kapalika, the old brethren of the end? The social function of the Kapalika is only to know.

This does not sound so much, only to know, but to live in the awareness of the truth accomplishes dual functions. First, it slowly compels one to act differently, by degrees. Perhaps we say one tonne of carbon each is our real limit and then over ten years try to approach it. Perhaps we say each meal I eat from this bowl is one meal nearer becoming as dead as its donor and then try to live right, whatever that means by our lights. This individual function, to change what we live, to be in accordance with the truth that things end, is the fundamental satyagraha.

The second function, however, is the one where I want to throw you a bone of hope. The second function of the Kapalika is to strip away the lies about death, the mythology and the avoidance, and to spread hope by a simple fact: the avoidance of the truth of death is worse than death itself. Death cannot be avoided, but its avoidance can be avoided.

Over-consumption and aggression to the planet and other people cannot go on indefinitely, and we will either transform or crash, but the age of the mall-dinosaurs is over and living in the truth strips out the lies. It does so without any bold statement, with no advert in the papers proclaiming that the end is nigh, but with the gentle and gradualist refusal to acknowledge other people's social fictions around consumption. It is the least we can do.

Reincarnation is the fundamental doctrine of the East, present in Hinduism, in Buddhism, and in the background of Taoism. The end of life joins to the beginning, and it's all a single light, death-orgasm-birth, a single psycho-mortosexual moment, the timelessness which is The Beyond. We live in a culture which has made birth relatively safe, sex less mysterious, and death largely invisible, yet still from all three timeless points, the numinous shines.

So too, there is a strange numinosity around the death of capitalism, the survival challenges we pose the ecosystem, and the green shoots of a new culture which ache to climb the wreckage, and instead find themselves shadowed by dead-while-standing oaks. We sense in it the birth of a new world, not the quiet progression of the new better replacing yesterday's best, but the wracking collapse of everything we have known, and a rebirth, with all the pain and trauma and blood which goes with that sacred mystery.

Things fall apart; the centre cannot hold. In fact, nothing can. To wish for collapse is as foolish as to wish for our own early deaths to see new life. To seek to postpone it may be as rational as eating healthily, or as irrational as a fourth heart transplant. To remain conscious that we seem to be taking a turn much for the worse, and to make sensible preparations, to put our affairs in order, to make a will, and to tell our friends and relatives that we love them, is a sane and sensible way to face the end.

Go to Paris one last time. Enjoy the steak. As you bite off and chew these experiences of the outgoing global order, consuming a little of the death of the world, taste it fully, this life of unbridled excess and borrowing against the accounts of future generations.

It tastes good, regardless of what it means.

And then, one day, in awareness, the bitterness behind the sweetness can be tasted, and we lose all desire to live by the suffering of others, and honest, non-destructive labour becomes enough.

But you will not cheat nature, and we cannot awaken others on our own schedule.

So you live your life in truth, day by day, every meal from the skull bowl, wondering what else there is to learn from it.

ROBERT MONTGOMERY – Ostend Wall piece, 2016 – *Issue 10. Autumn 2016*

Photograph courtesy of the artist and Crystal Ship Festival

Mountaineers

As the manifesto made its way into the world, Dark Mountain became a gathering point for a network of people to whom the stories of our civilisation no longer made sense. Among them were the writers, thinkers and artists whose work you met in this book.

Hardly any of its contributors knew each other a decade ago, but there are some here who have become close collaborators and friends in that time, some who came along out of curiosity and ended up making themselves indispensable, some with whom we shared nights around a campfire or conversations that changed the way each other see the world - and some who remain names over email, with clues gleaned from each other's words and images or from short biographical notes like those that follow.

Akshay Ahuja is a writer and editor, and blogs, occasionally, at the Occasional Review. As well as contributing to several issues, he reviews books for the Dark Mountain blog. He lives with his wife and son in Cincinnati, Ohio.

Jason Benton has been following a rabbit trail that began with woods and fields and creek-bottomed hills, then cities and oceans, airplanes and theology, mountains and jungles, horses and psychiatric nursing, and now goats and gardens. Presently, he is back home in Western Pennsylvania, with children and homeschooling, work, a very small farm and a large garden.

Nancy Campbell is a writer and book artist whose work responds to polar and marine environments. Her books include *How To Say 'I Love You' In Greenlandic: An Arctic Alphabet* (winner of the Birgit Skiöld Award) and the poetry collection *Disko Bay*, shortlisted for the Forward Prize for Best First Collection 2016.

Florence Caplow is a Unitarian Universalist minister, Soto Zen priest, former conservation biologist, activist, writer, and editor. She edited and contributed to *Wildbranch: An Anthology, of Nature, Environmental, and Place-Based Writing*, and *The Hidden Lamp: Stories From Twenty-Five Centuries of Awakened Women*. Her essays can also be read online at Slipping Glimpser: Zen Wanderings and Wonderings.

Cate Chapman works as co-director of the Ecological Land Cooperative alongside managing a small copy-editing business, Skylark Editing. After stumbling across the Dark Mountain Project online one dark night and getting thoroughly overexcited, she has contributed to several books and recently joined the editorial team. When not hunched over a laptop, Cate can be found in the woods, incompetently carving spoons.

Warren Draper makes pictures out of words, pixels, paint and plants. He walks the forgotten places of South Yorkshire with seeds in his pocket and mud on his mind. Together with his fellow artist Rachel Horne, he created a Dark Mountain-inspired arts festival, *The Telling*, and arts and culture magazine, *Doncopolitan*, which set out to prove that there is no such thing as a cultural desert. horneanddraper.com

Charlotte Du Cann writes about mythology, metaphysics and cultural change. She used to write about the consumer society until she sold everything she had and went on the road - a journey she charted in *52 Flowers That Shook My World - A Radical Return to Earth* and other books. In 2011, she reported on the Uncivilisation festival for the *Independent* and offered to help out behind the scenes, soon becoming a core part of the Dark Mountain team. charlotteducann.blogspot.co.uk

Tim Fox writes non-fiction that is centred on the question of ecological integrity. What does it mean? And how might it be realised and preserved, in this time of worldwide ecosystemic disintegration? His writing can be found in the anthology *Forest Under Story*, Orion magazine, the Yes! Magazine website and online publication The Forest Log, as well as Issues 4, 5, 9 and 11 of Dark Mountain. He draws particular inspiration from 30 years of immersion in the central Oregon Cascades where he lives with his family.

Kim Goldberg was a freelance journalist and non-fiction author for 20 years, writing about the environment and social justice. Then she became immersed in Chinese martial arts and, for a decade, she could not write anything beyond a grocery list. When she awoke from this, she found an ocean of poems tumbling out of her and it hasn't stopped since. She learned of Dark Mountain from another contributor, Patricia Robertson, and knew immediately that she wanted to be part of it. She lives on Vancouver Island where, during the herring spawn, she can hear gulls crying and sea lions barking all night long.

John Michael Greer is the author of The Archdruid Report, a weekly blog on peak oil andcurrent affairs, as well as more than forty books on subjects ranging from nature spiritualityto the future of industrial society. He lives in an old mill town in the Appalachians with his wife Sara.

Vinay Gupta is part engineer, part mystic, descended from Scottish peasants and Indian scholars. He is the inventor of the Hexayurt, a simple open source shelter, favoured by Burning Man campers and the Pentagon. He met Dougald Hine in a squat in London in early 2009, a story that is told in 'Black Elephants and Skull Jackets', published in the first issue of Dark Mountain.

Ron Hagg is an exhibited photographer, based in Taos, New Mexico, and has worked in the field of education for most of his adult life. His first job was helping migrant farm labour families. When he worked at Hoopa Valley High School in Humboldt County, California he initiated and coordinated the effort to teach the languages of three of the area's Native American tribes, Karuk, Yurok and Hupa. He has written four novels: *Jesus of Kneeland*, *To Keep From Drowning*, *Dreams of a New Day* and *Escape In Time*. The last of these was turned into a full-length motion picture. haggmedia.com

Dougald Hine read a blogpost by a guy called Paul Kingsnorth, floating an idea for a 'deeply, darkly unfashionable and defiant' new publication, dedicated to wildness and beauty, and thought this sounded like something that ought to happen. Out of the conversations that followed, they wrote *Uncivilisation: The Dark Mountain Manifesto*. These days, he lives in a city by a lake, an hour's train ride west of Stockholm, works with Riksteatern, the Swedish national theatre, and writes about how we can disentangle our thinking and our hopes from the cultural logic of progress.

Ian Hill grew up in the east of England, but has lived in Cumbria for 25 years. His writing focuses on landscape and memory, and has appeared in Earthlines magazine, in two issues of Dark Mountain, and in *The Language of Footprints*. He was the recipient of a Hunter Davies Bursary for new writers in Cumbria in 2012. A selection of his work is also on his blog, which was shortlisted for the Blog North awards in 2013. printedland.blogspot.com

Bruce Hooke is a photographer and performance artist residing in the small, western Massachusetts town of Plainfield. His work focuses on the evolving human relationship to nature. He heard about Dark Mountain from a friend and felt an immediate connection between his work and the mission of the project. Bghooke.com

Nick Hunt is a writer and storyteller. His first book *Walking the Woods and the Water* is an account of a long walk from the Hook of Holland to Istanbul. His second book *Where the Wild Winds Are* is the story of following Europe's winds. He also writes fiction. Attempts to understand the myths that lie beneath collapse and change led him up the Dark Mountain, where he now works as an editor, contributor and friend. Nickhuntscrutiny.com

The late **Glyn Hughes** was a poet, novelist and an early supporter of Dark Mountain. For forty years, he lived in and wrote about Calderdale, West Yorkshire, and his book launches would pack out local venues, but his work also received recognition on a national scale: two of his books were selected as 'great classics of British nature writing' by readers of the *Guardian*, while the *Times* called him one of 'the best authors ever on the north of England'. His interview with Paul Kingsnorth for Issue 2 of Dark Mountain would prove to be the last before his death from cancer at the age of 75 in June 2011.

Garrett Hupe is a resident of the American Midwest who has been moun-taineering for about seven years. The title of the image, 'Where to? Where from?', invokes the question of the future, the question of the past and so the question of the present. The mountain, its reflection and the darkness in between invoke this formula, itself an approximation of the Sanskrit, TATHĀGATA, a word with dual meaning, either 'Thus Gone or 'Thus Come'.

Nicholas Kahn and **Richard Selesnick** are a collaborative artist team who have been working together since they met while attending art school at Washington University in St. Louis in the early 1980s. Both were born in 1964, in New York City and London respectively. They work primarily in the fields of photography and installation art, specializing in fictitious histories set in the past or future. They have participated in over 100 solo and group exhibitions worldwide and have work in over 20 collections, including the Brooklyn Museum of Art, the Philadelphia Museum of Art, the Houston Museum of Art, the Los Angeles County Museum of Art and the Smithsonian Institution. In addition, they have published 3 books, *Scotlandfuturebog*, *City of Salt*, and *Apollo Prophecies*.

Thomas Keyes lives, gardens and forages in the highlands of Scotland. In the birch woods and the roadkill deer, he has found both raw materials and subject matter. Having started out as a graffiti artist in Belfast, his most recent work fuses the visual language of street art with the traditions of Celtic and Pictish illumination. Thomaskeyes.co.uk

Paul Kingsnorth is co-founder of Dark Mountain. He is also the author of two novels, *The Wake* and *Beast*, two books of non-fiction, *One No, Many Yeses* and *Real England*, and one collection of poetry, *Kidland*.

Robert Leaver is currently living in a state of shock. But he continues to function, more or less, as a husband, father, friend, writer, artist and musician. He he based in NYC and the Catskill Mountains. He can't quite recall how he came to Dark Mountain, but has never been more grateful for its existence. More on his activities can be found at robertoleaver.com

Emily Laurens is one of Feral Theatre's co-founders and co-directors and lives and works in rural West Wales. Her work uses and utilises myth, clown, ritual, mask, puppets and movement to explore loss, transformation and our connection to the Earth and its non-human inhabitants.

Hannah Lewis is founder of the Remakery in Brixton, South London, a shared workspace for reuse and upcycling enterprises. She is also a lover of language, but has approached writing with caution since university, when the breakdown of her worldview shattered her belief in the possibility of articulating anything meaningful. Dark Mountain's search for stories appropriate to our time helped revive her courage to write.

Sylvia V. Linsteadt is a writer and animal tracker living in the hills of Northern California, where she was born. Her first novel *Tatterdemalion*, a collaboration with artist Rima Staines, was published by Unbound in May 2017. Much of her work is set in a future California, where old stories have come alive again. She discovered Dark Mountain through Rima, already a mountaineer and painter of covers, who mentioned the name to her one day five years ago. Reading the manifesto was an act of coming home, and she has been sending her work to those wise and shadowed peaks ever since.

Marne Lucas is a multi-disciplinary artist based in New York City. Using photography, video, sculpture and installation to explore nature, culture and sexuality, she presents social, aesthetic and eco-based philosophies about humanity. Her new project 'Bardo' is a collaboration with hospice patients. She recently participated in an Arts/Industry residency at the Kohler Co. factory. marnelucas.com

Robert Montgomery follows a tradition of conceptual art and stands out by bringing a poetic voice to the discourse of text art. He creates billboard poems, light pieces, fire poems, woodcut and watercolours. He was the British artist selected for Kochi-Muziris Biennale 2012, the first biennale in India. He has had solo exhibitions at venues in Europe and in Asia, including major light installations at the old US Airforce Base at Tempelhof. The first monograph of his work was published in 2012 by Distanz, Berlin. robertmontgomery.org

Kim Moore's first collection *The Art of Falling* was published by Seren in 2015. Her poem 'In That Year' was shortlisted for the 2015 Forward Prize for Best Published Poem. Her pamphlet *If We Could Speak Like Wolves* was a winner in the 2012 Poetry Business Pamphlet Competition.

Narendra's association with tribal communities began when he undertook field-research from 1980 to 1985 in the deeply interior Abujhmad region of Bastar on a United Nations University project, Tribal Perceptions and the Modern World. Though the project formally concluded in 1985, he continued his field study till 2013 on issues of adivasi ecological-cultural expressions, notions of forests and wilds, ecologically-inspired modes and institutions of governance, adivasi survival in a market economy, ecology and folk, knowledge and learning. Along with two other friends in Bastar he ran an unstructured and informal initiative, DoE (Dialogue from the Other End). It was an attempt to help revitalise the disappearing conversations on land, water, forest and socio-ecological cohesion in everyday living that sustained the unique Adivasi way of life.

Gregory Norminton was born in Berkshire in the middle of the 1976 heatwave. As a north European mongrel, he was genetically designed for a cooler planet than the one we are creating. A novelist, translator and university lecturer, his books include *Ghost Portrait*, *Serious Things*, *The Devil's Highway* and a collection of aphorisms, *The Lost Art of Losing*. He lives with his family in Sheffield.

Jacob Pander is a filmmaker and graphic novelist. He wrote and directed the award-winning feature film, *Selfless* and the documentary film *Painted*

Life. He has also directed music videos for Fantastic Plastic Machine, Howie B. and Spacer, and shorts including Subtext. Conceptual work in collaboration with Marne Lucas includes the infrared video installation Incident Energy and cult classic short film THE OPERATION. jacobpander.com

Persephone Pearl is an artist and producer who uses colourful, accessible work to approach big questions. She co-founded Feral Theatre and Remembrance Day for Lost Species, and is co-director of ONCA in Brighton. She was drawn to Dark Mountain as a space that welcomes experimentation, grief and the experience of paradox.

Lionel Playford works with landscapes of many kinds, exploring through drawing and other media what it is to viscerally experience contemporary landscape as a person living in a technologically advanced society. His landscapes range from the post-industrial areas of late 20th century northeast England to peat moorlands of northern England and Finland. His contribution to Dark Mountain came about through a chance conversation in his studio with Nick Hunt who had come to Garrigill in search of firewood.

Sarah Rea credits Dark Mountain with ending a 10-year writer's block. After reading an article in the *New York Times* and devouring the manifesto, she found that the deadline for submissions for Issue 6 was only days away. Fresh out of a long and tumultuous relationship, living in a tiny room in a cold cabin with her hunting dog, she wrote her piece, sent it in and almost forgot about it. But the piece was accepted for publication - and three years later, Sarah is writing full-time as a reporter for a small newspaper in her mountain home of Mammoth Lakes, California. She also plays mandolin in a band with her sister, they call themselves The Daughters Rea and sing about things both dark and bright.

John Rember lives and works in the Sawtooth Valley of Idaho. He is the author of five books: *Coyote in the Mountains*, *Cheerleaders from Gomorrah*, *Traplines*, *MFA in a Box*, and *Sudden Death, Over Time*. His contribution to this anthology is taken from a yet-to-be-published manuscript, *A Hundred Little Pieces on the End of the World*.

Eric Robertson was googling for queer ecology, during graduate studies in environmental humanities at the University of Utah, when the quirks of the internet led him to Paul Kingsnorth's essay, 'Dark Ecology'. As a gay man living in Utah, surrounded by hordes of future-obsessed, baby-making Mormons, the idea of 'the progress trap' lit his hair on fire. He now explores how nonreproductive humans can withdraw, creatively and purposefully, from cultures bent on eternal increase.

David Schuman grew up in a New Jersey suburb about thirty miles from Manhattan. His parents, both teachers, told him he could do anything he wanted in life as long as it wasn't teaching. So, of course, that is what he does now. He lives in St Louis, at the nexus of many aspects of America's present cultural moment. Discovering Dark Mountain through the *New York Times*, he found the story 'Squirrel', which he had written months earlier, made sense in a way that it hadn't before. So he submitted it for publication.

Martin Shaw is a writer, teacher, mythologist and good friend to Dark Mountain. At the project's festivals and gatherings, crowds are held in the magic of his storytelling. He runs the Westcountry School of Myth, the creator of the oral tradition course at Stanford University and the author of *A Branch from the Lightning Tree*, *Snowy Tower* and *Scatterlings: Getting Claimed in an Age of Amnesia*.

Tom Smith is a PhD candidate at the University of St Andrews and a member of the Dark Mountain editorial collective. He's interested in environmental philosophy, radical geography, and pulling on whichever threads unsettle civilised thought about what it is to be human.

Maria Theresa Stadtmueller lives on a permaculture-y farm in Vermont where she scythes a lot. She creates The Big Chew Podcast, about living a new, science-based Earth story. She got an MFA in writing from The University of Iowa, and was a comedian in New York. A friend turned her on to Dark Mountain, a refuge of sanity.

Carla Stang is an anthropologist (PhD. University of Cambridge) who has always been fascinated by how different people actually experience the world, and especially the land. She has written a book about living in a Mehinaku village called *A Walk to the River in Amazonia* (2009). Currently she is co director of studies of the M.Phil. programme at Schumacher College.

Em Strang is a poet, editor and prison tutor. From Issue 6, she was Dark Mountain's poetry editor, a role which culminated in a special edition,

Uncivilised Poetics, Issue 10. Her illustrated pamphlet, *Stone* (a collaboration with artist Mat Osmond) was published in March 2016 by Atlantic Press and her first collection *Bird-Woman* in autumn 2016 by Shearsman.

Anita Sullivan has lived in the Pacific Northwest for 34 years, but her heart remains in the high desert country of New Mexico, near the sky-eating mountains. For the last 14 years of his life, she was married to a retired religious studies professor who translated the Book of Job and the Song of Songs from the Hebrew Bible. They had long conversations about poetry, stories and music. She found her way to Dark Mountain through her brother, who follows all sorts of websites and blogs that deal with the state of the planet. Born under the sign of Libra, this may explain why she became a piano tuner, since tuning is all about balance.

Kate Walters' art and writing is inspired by being in wilderness. She works with a technique called 'becoming the hollow bone', an ancient tool which allows her to reach invisible worlds. Her watercolours and drawings have been shown all over the UK and she is soon to embark on projects with ONCA Gallery in Brighton and Tremenheere Gallery in Cornwall. katewalters.co.uk

Steve Wheeler is a writer, editor and performer based in the South-West of England. He is one of the editors of the Dark Mountain books, and his own work has been featured in several issues. He also runs self-rewilding workshops exploring the connections between physical, emotional and cultural freedom, and is a practising acupuncturist, bodyworker and natural health consultant. @steel_weaver

A Message from
the Dark Mountain Project

Everything you've read in this collection (and much more beside it) first appeared in the pages of the Dark Mountain books. Since the Project began, those books have been made possible by the generosity of our readers.

So if this introduction has left you hungry for more, please consider taking out a subscription to future issues of Dark Mountain. As a subscriber, you get each new book as soon as it comes out for less than the usual price.

You'll also find many of our earlier issues are still in print and available through our online shop.

For more details, visit:
www.dark-mountain.net

CPSIA information can be obtained
at www.ICGtesting.com
Printed in the USA
FFOW03n0503290317
33943FF

9 781603 587419